101 STORIES *behind the* ORIGIN *of* EVERYDAY THINGS

Seema Gupta

PUSTAK MAHAL®

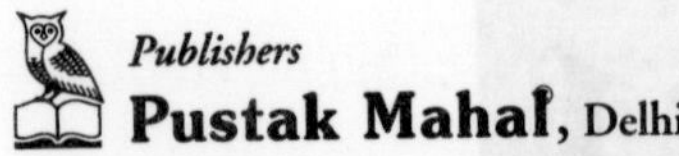

Publishers
Pustak Mahal, Delhi

Administrative office and sales Centres
J-3/16 , Daryaganj, New Delhi-110002
☎ 011-23276539, 23272783, 23272784, 23260518
E-mail: info@pustakmahal.com • *Website:* www.pustakmahal.com

Branches
Bangalore: ☎ 080-22234025, 40912845
E-mail: pustakmahalblr@gmail.com
Mumbai: ☎ 022-22010941, 22053387
E-mail: unicornbooksmumbai@gmail.com

ISBN 978-81-223-0909-6

Edition : 2020

This book was earlier printed under the title—
"Origin of 101 Everyday Things"

Printed at : Unique Color Carton, Delhi

Preface

There are so many things lying scattered around us that we use everyday without giving a second thought about their origins. We are so used to them that we feel they have always been around. But is this true? Have you ever wondered, where have these things come from?

Each and every thing in this world has a tale of its own regarding its origin. Consider the pen in your hand, or the television that entertains you. These things did not exist barely a few decades or a few centuries ago.

The eraser, pencil, shoes, chairs, tables and so many other things in our lives are the successful outcome of research being conducted over the centuries. The hard work and dedication of these inventors and discoverers has gone into making these things. Scientists and inventors worked day and night, even labouring under unfavourable conditions, to bring these things into our lives.

Today we consider ancient or medieval times as being backward because then there were no electronic goods, while today there are various machines, gadgets and gizmos. But could all this have been possible without the efforts of those people from the ancient era? Absolutely not!

So we have to continue this tradition of inventing and discovering new things by keeping our eyes and ears open and also by learning from the past experiences of these great men and understanding their way of working.

Our book, *Origin of 101 Everyday Things*, delves into the invention, discovery or making of 101 select things which you use everyday. This book differs from others of its kind as it not only elaborates on how these things originated but also narrates some interesting anecdotes connected with these things – something you will not find in most other books. This is what sets this book apart from other me-too books.

The colourful display of pictures and the swift narrative will hold the interest all age groups, transporting readers into an era of inventions and discoveries and wonderment.

Contents

1 Aeroplane

I have no feathers, I have no beak
Still I fly high across the peak
...Can you guess my name?

From times immemorial, man wanted to fly in the sky like a bird. Some people tried to fly by tying wings onto their hands; others tried jumping from mountaintops with wings tied to their back. Many were injured, some died but no one succeeded in their endeavour. However, all these unsuccessful experiments made one thing absolutely clear – without a mechanical instrument it was impossible to fly.

Considering this fact, George Cayley of England built the world's first glider in 1804. Fifty years after the invention of this glider, in 1853, Cayley built another glider that could carry one person. He sent his driver as the first pilot of this glider. As the glider went past a deep valley, the poor man almost had a heart attack. When his feet touched the ground, he thanked his stars and ran for his life. He even resigned from his job with Cayley!

By the end of the 19th century, many others had built gliders but they could not travel a long distance. The two Wright brothers, Orville Wright and Wilbur Wright, from Ohio City tried to build a glider that would travel long distances. They sold all their belongings and invested all their savings in this machine. They even fitted an engine of 12 horsepower in the glider.

Orville Wright took off in this glider on 17th December from Kitty Hawk near Northern Carolina. The flight went up to 120 feet, much higher than previous gliders. But this flight could not attract much attention because people did not see much utility in such a flyer! Yes, they had named this glider 'Flyer'.

However, the Wright brothers did not lose heart. They improved the first model of the flyer and built a much better version in 1907, which could fly for 24 miles at a stretch. Now this was something. The media extended good publicity to this flyer. The then President of America, Theodore Roosevelt instructed his armed forces to include these flyers in their war equipment.

There have been many amusing incidents in the history of flying. Initially, when these flyers flew at 30 miles per hour, passengers played table tennis on the roof of the flyer! When the postal air service was started, during its first flight from Washington to New York, the pilot lost his way and reached Maryland instead! Afterwards the letters were sent to New York by train!

Today, such mishaps are unthinkable. Technologically advanced radar systems, high-flying jets and supersonic aircrafts can tread the skies at speeds of 4,500 miles per hour and are indeed a dream come true.

2 Air-conditioners

During the 18th century, the Khalif of Baghdad, Al Mehndi, lived comfortably in his royal palace. He had many servants at his beck and call. The Khalif had just one problem in life. Summers were unbearable for him due to the heat. One day, he thought that if he made the walls of his room double their size and filled ice between them, then maybe the summer would become bearable!

The method proved to be a success. But it was quite tedious, as the servants had to fill the gap with ice every day. Since it was not possible for a common man to afford such luxury, this primitive method of air-conditioning was limited to the Khalif.

For many years thereafter, man used natural resources to keep cool in summers – like building thick walls, high roofs, covering the roof with extra layers etc. Modern air-conditioning only began about 150 years ago.

Dr John Gorrie of Florida had many patients suffering from malaria and yellow fever. The terrible heat added to their woes. In 1851, to overcome this problem, Dr Gorrie built an air-compressing machine, which sent cool air when passed over ice cubes. It was around this time that Ferdinand Carre of France built the world's first ammonia coil, which could suck the heat and humidity from air.

In 1902, the air compressor and the ammonia coil were joined together to build the world's first air-conditioner. Mr Willis H. Carrier built it and it was installed in Brooklyn

Printing Company. Slowly, the techniques and designs of air-conditioners improved.

Today air-conditioners are available in all sizes. They don't just cool a room but have the capacity to cool an entire building. There are huge air-conditioning plants that serve a dual purpose: they are used for cooling in summers and for heating in winters!

The world's largest air-conditioner is installed at the World Trade Centre in New York. Its compressor and pump are spread over an area of 2.5 acres and the pipes are 170 m long.

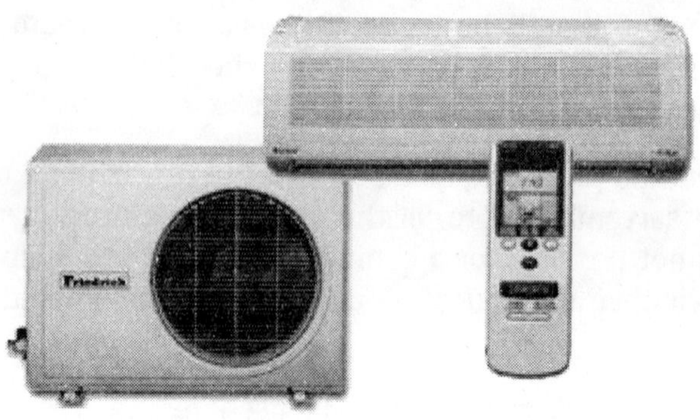

3 Ambulance

The ambulance has given life to many dying patients with its quick service. Its story goes back to 1792 when Napoleon's personal physician Baron Dominique Jean Larrey had the brilliant idea that wounded soldiers should be taken out of the battlefield on a horse carriage.

Before then, there was no provision to carry them out and wounded soldiers were usually left to their fate.

Dr Dominique, together with the chief surgeon of the French army, Dr Pierre Francois Percy, established the Ambulance Corps. Each division had 12 such ambulances. This corps was used in the 1796 war between France and Italy. It was a great success. Many countries followed suit and built their own fleet of ambulances.

In 1864, the ambulance was officially accepted in the International Geneva Agreement and it was decided that no one would attack an ambulance during war.

In 1870, France experimented with balloon ambulances. Wounded soldiers were put in these hot air balloons and sent high up in the air hoping they would reach some safer place! But this was not a reliable method. Soon their use was discontinued.

With the invention of motorcars, the ambulance also shifted focus. In July 1900, the army built its first ambulance car. Now ambulances have become more comfortable and spacious and are even fitted with life-saving equipments.

4 Anaesthesia

In 1842, Dr Crawford Long of Jefferson city in America had to face public fury when people demonstrated outside his clinic and threatened to kill him. Do you know why? Dr Long used to operate on his patients after giving them a dose of Ether, so that they would not feel pain. A young student James Venable was his first patient, whom he operated after administering ether. Venable felt no pain and the smell of ether had no ill effect on his health either.

The word spread soon and some mischievous young boys requested Dr Long to let them smell ether for sheer pleasure. Sometimes, Dr Long would accede to their request. However, he operated upon nine patients successfully. He would charge $2.25 for ether separately. The older generation in the city did not like this idea of paying for unconsciousness! They accused Dr Long of using voodoo (black magic) and raised an outcry against him. Dr Long was helpless. He gave up using ether and that was the end of this story.

Almost four and a half years later on 16 October 1846, the use of ether was demonstrated by Dr John Collins Warren at a hospital in Massachusetts before a panel of doctors. Dr Warren operated upon a 20-year-old young man Gilbert Abbott and removed a tumour from his jaw. When Abbott gained consciousness, the doctors asked him if he felt the pain. Abbott denied having felt any pain at all. This was the beginning of a new era in the world of medicine.

Before the discovery of anaesthesia, a surgeon's job was thought to be merciless. People would shy away from doctors, as they dreaded painful operations. At some places, patients were drugged with opium, wine, alcohol etc. so that they would not feel the pain. Some doctors used hypnotism also.

But the discovery of anaesthesia changed all this. And it is with the help of anaesthesia alone that medical science has managed to progress so much.

5 Antiseptic

In ancient times, open wounds were treated with human urine. Another way to cure a wound was to apply alcohol mixed with vinegar. In ancient India, the wound was cleaned with hot water and then honey and sesame oil was applied on it.

The people of ancient Egypt were far ahead in treating wounds. They had discovered certain chemicals, spices and oils that were used to preserve dead bodies. Similar chemicals were used to treat wounds.

When the microscope was invented in the 16th century, doctors thereafter discovered micro-organisms like bacteria and viruses. Till then no one knew that these organisms were responsible for decaying of wounds.

In the 19th century, French scientist Louis Pasteur discovered this fact. In the meantime, English surgeon Joseph Lister contemplated looking for some ways to kill these micro-organisms, as many of his patients had to suffer severe complications due to decaying of wounds. Lister started washing his tools, hands and patients' wounds with phenol or carbolic acid. This proved quite effective. Encouraged by the results, Lister sprayed phenol in his operation theatre.

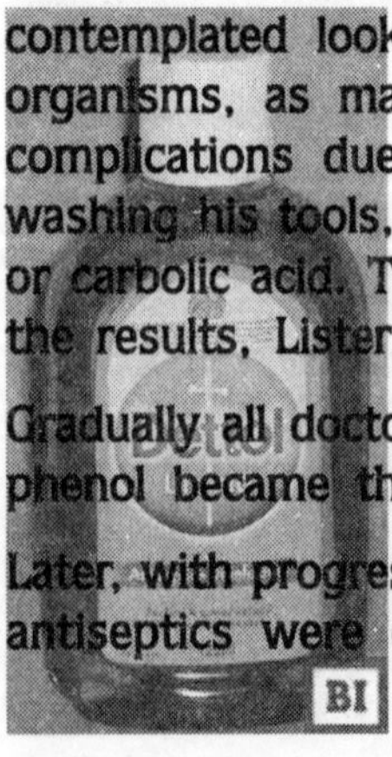

Gradually all doctors in England started using phenol. And phenol became the world's first antiseptic.

Later, with progress in medical science, other more effective antiseptics were discovered.

6 Aspirin

I am an analgesic, antipyretic.
I can even prevent a heart attack.
I am the magic medicine.
Yes, I am the amazing aspirin.

Aspirin is the most common of all pain relievers. The active ingredient of this pain reliever, Acetyl Salicylic Acid, has been known to man for at least 2000 years. It is present in the bark of the willow tree and the silver birch tree. But it was previously used in its natural form.

In 1763, an English doctor Edward Stone began to study the properties of these plant substances and prescribed its juices as pain relievers. In 1853, Carl Frederick Gerhardt first artificially synthesised this ingredient in France. But his experiments did not go too far.

In 1890, Felix Hoffmann was employed with the famous German pharmaceutical company Bayer as a scientist. Unfortunately, his father was suffering from arthritis, which was very bad and quite painful. His cries made young Hoffmann restless. One day, he decided to prepare a medicine that could relieve his pain.

In 1897, after intense labour of seven years, Hoffmann managed to prepare acetyl salicylic acid. He took the sample to many doctors. When they tried it on their patients, the results were outstanding. Hoffmann was elated. He managed to relieve not only his father's pain but that of millions of people all over the world too.

Though very effective, the medicine still took 18 years to hit the market. In 1915, the Bayer Company started selling this medicine in a pack of 20 tablets each.

Acetyl salicylic acid is prepared from the chemical salicin found in the leaves, bark and roots of the willow and meadow trees, which grow in marshy areas. It is said that wherever a disease breeds, its natural medicine would be nearby. Since fever and body ache are common in marshy areas, scientists experimented on the natural vegetation around and found this miraculous cure.

Now, the question remains: why is this medicine called *aspirin*. It is believed that the word *aspirin* has been derived from the word *Spirea* – the scientific name of the meadow tree. The Bayer Company of Germany used aspirin as a brand name, which has stuck till date. Today aspirin worth $7 billion is prepared every year all over the world.

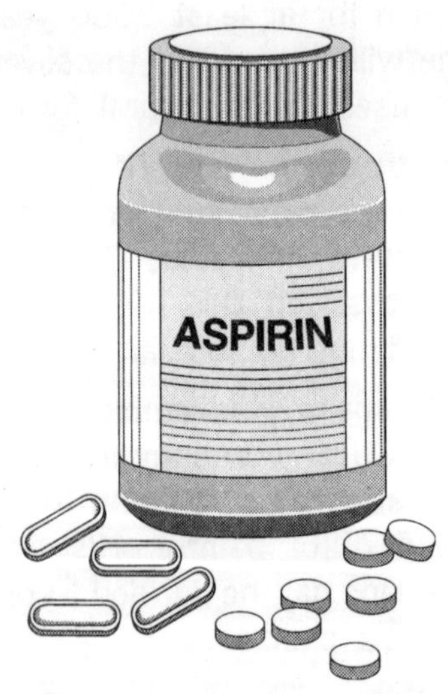

7 Ball Bearings

The world's first ball bearing was made in France in 1545. It so happened that the famous French sculptor Florentine built a huge sculpture for the Emperor. The figure was so heavy that it could not be moved easily. A novel idea then struck Florentine. He thought of fitting the base of the sculpture with tiny round balls made of solid gold on which the sculpture could slide easily. He called upon Benvenuto Cellini, the goldsmith, and asked him to prepare these tiny balls. This is how the world's first ball bearings came into existence.

To be fair, though, the credit for the ball bearing should go to Leonardo da Vinci, who had already drawn its sketch in his notebook.

Ball bearings were not much in demand before the Industrial Revolution. In 1679, after the advent of the Industrial Revolution, Philip Vaughan made iron ball bearings that are fitted on a wheel.

Then came the era of the bicycle. In 1869, James Moore of England had ball bearings fitted in his bicycle. When he entered the cycle race on this bike, he won with flying colours. Soon, ball bearings became one of the essential accessories in cycles.

In 1930, English designer Frederick Lanchester used roller bearings in the gearbox of a motorcar and now ball bearings and roller bearings have become essential accessories of all motor parts. Today it is impossible to imagine the world without ball bearings.

8 Balloon

Kids love me. My attractive colours and shapes enchant them. Yes, I am their favourite balloon.

The pages of history talk about balloons being made with paper in China around 200 BC. The ancient Japanese art of origami also boasts of paper balloons. But these balloons would not float in air.

In 1643, Italian scientist Torricelli hypothesised that air is not just nothing but a mixture of many gases. Then the idea struck many people that if air were sucked out from the paper balloon then surely it would float in air. On second thoughts, filling these balloons with a gas lighter than air would also serve a similar purpose. So balloons were filled with hot air. Soon small hot air balloons were a major hit with little children.

In 1783, French scientist J.A. Charles discovered hydrogen gas, which was lighter than air. Now taffeta balloons filled with hydrogen were being sold in the market.

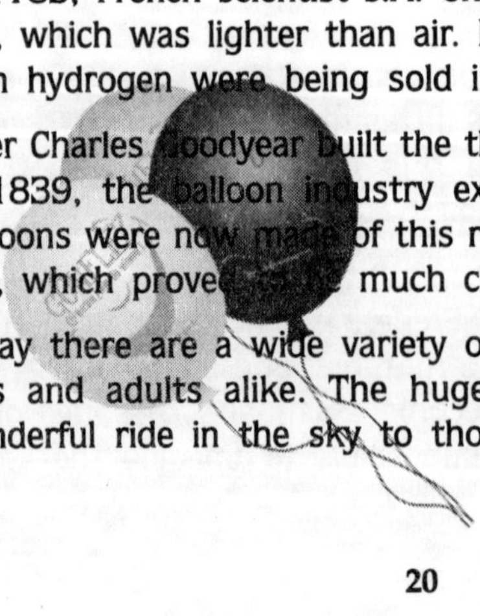

After Charles Goodyear built the thin, strong, flexible rubber in 1839, the balloon industry experienced a major boost. Balloons were now made of this rubber filled with hydrogen gas, which proved to be much cheaper than taffeta.

Today there are a wide variety of latex balloons to amuse kids and adults alike. The huge hot air balloons give a wonderful ride in the sky to those seeking adventure.

9 Ballpoint Pen

I am not so ancient as a pencil or pen.
Give me a chance I shall be your friend.
.....Indeed, I am quite modern in my origin and looks.

In 1945, a long queue before a departmental store in New York attracted everyone's attention. People were coming out of the store in droves after buying a special pen costing $12.50. The store claimed this pen could write even underwater. As proof, the store had displayed a pen in its shopping window, which was shown writing in water.

Within a week, the store had sold 25,000 pens. These pens were nothing but today's Ballpoint Pens, also known as Dot Pens. Although he had not invented the ball pen, but all the advertising of the pen was the brainchild of Milton Reynolds of Chicago. It was Lasalo Biro of Argentina who had invented the ballpoint pen in 1937 and he had received its patent along with his brother George Biro too.

The most interesting part of the story is that, simultaneously, an adversary of Reynolds publicised a pen called the Rocket Pen. The advertisement for the Rocket Pen went like this: 'This pen can melt locks, write on stones, remove your grey hair and so on...' In reality, no such pen existed! The rocket pen could not influence customers of the ballpoint pen. But the ball pen's quality became its own enemy, when it started leaking frequently.

Within a few months, most customers who had earlier bought the pen with much enthusiasm came back to the store to return it. People felt the magic of the ballpoint pen waning. It was feared that this was the end of the ballpoint pen's saga. It was not to be.

The manufacturers improved the quality of the pen and also reduced its price. Soon the improved version of the ballpoint pen was ready. Now it was priced at merely 19 cents. Today ballpoint pens are used all over the world. Millions of pens are sold in the market everyday.

10

Baskets, Bags and Suitcases

These are three items that serve a similar purpose. Which was discovered first it is difficult to say. Read on...

Baskets and bags were invented almost at the same time. Men needed bags to carry their things, while women required baskets to keep their belongings. Ancient Greece and Rome had beautifully designed handmade baskets. There were bags also, but not shoulder bags, since people used handbags in those times.

In the Middle Ages, women also started using handbags. What began as a useful thing while shopping soon became a fashion statement. No woman could be seen without a handbag now. Leather handbags and handbags made of cloth were used at that time. Cloth handbags were decorated with intricate, fine embroidery.

The story of suitcases began in the 19th century. It is said that during the reign of Queen Victoria of England, her Prime Minister W.E. Gladstone was the first person to use a leather suitcase. Gladstone travelled extensively. So he ordered a leather suitcase to suit his needs. This suitcase had different compartments – one for clean clothes and another one for dirty clothes. With the successful use of the suitcase, big trunks were also made. It became a fashion to travel with these big trunks that helped one carry all necessary household items, including clothes.

Famous Hollywood star Elizabeth Taylor carried 10 huge trunks whenever she travelled. On one of her trips, she had a total 128 suitcases and bags!

With the beginning of air travel, it became difficult to carry such heavy luggage. Now lighter and smaller bags entered the market, which were called airbags. Airbags were spacious and light too. Today lightweight suitcases are also in the market and serve the purpose of a trunk by loading in lots of stuff. Briefcases are a smaller version of suitcases. Business people use them mostly to carry their documents. Today, no executive or businessman can be seen without a briefcase and no woman can be seen without a handbag or shoulder bag.

That's how the bag journeyed from being a handbag to a shoulder bag and finally to an airbag and the suitcase journeyed from being a suitcase to a briefcase!

11 Battery

Hi! I am a battery. I am a source of energy. Famous Italian scientist Alessandro Volta is my father. The story of my birth goes like this...

In 1786, famous Italian scientist Luigi Galvani was conducting some experiments on dissected frogs. Suddenly he noticed that when a frog's leg touched a forceps or scissor, it was pulled back instantly. He tried it many times, with the same result. Galvani concluded that some sort of current was produced when a metal touched the frog's leg.

When Galvani's friend and famous physicist Alessandro Volta heard this, he tried a similar experiment with different metals. The results of his experiments indicated that some metals produce electricity. To prove his point, Volta built many plates of copper and zinc and placed them together on top of each other. Between each plate, he placed a cloth or a piece of cardboard dipped in saline water – this produced electricity. Napoleon Bonaparte witnessed this experiment when Volta demonstrated it before the National Scientific Organisation of France. This was the first battery of the world.

Even today the basic principle for making the battery remains the same. The unit of battery is the cell. The dry cells that are used in torches, transistors and the like are not really dry. They are filled with thick liquid.

Today batteries come in different shapes and sizes. The smallest are the button cells, which are used in watches and calculators. They weigh one-twentieth of an ounce. On the other hand, the world's largest batteries are used in submarines. They weigh almost a ton and are in the shape of a table.

There are many fish that produce electricity – like the electric ray, torpedo, electric eel etc. They are fitted with a natural battery that is recharged on its own.

12 Bicycle

Hi! I am a bicycle. I have two wheels. I don't need any fuel. I am an environment friendly vehicle. This is how I came into existence...

Baron Van Drais of Germany built the world's first bicycle. He had joined the two wheels with a wooden bar. To move these wheels, the rider had to kick the ground with force using his feet since there were no pedals. Baron called it Draisine. He also fitted a handle on the front wheel to move it sideways. This cycle was very expensive, so it was also called the Dandy Horse. Not many people were interested in it.

Later, in 1865, Michaux Lallement of France built the pedal-driven bicycle. The wheels were made of steel. But this bicycle was not very comfortable. Its ride was not smooth, so people named it the Boneshaker! It was around this time that someone suggested that if the front wheel was made bigger than the hind wheel, the bicycle would move faster. Such cycles were also experimented with, but they were very unstable.

Gradually, many changes were made in the bicycles. And then came air-filled tyres, which made cycling more comfortable. The two wheels were also made of equal size, the rider had a more comfortable seat and the wheels were moved with pedals. By 1885, these cycles hit the market and captured it in a big way. The modern cycle is an improved version of this model.

Today, cycles are much lighter and faster. Some even have gears just like in scooters and motorbikes.

13 Bottle

Hi! I am a bottle. Can you imagine life without me? From a newborn's milk bottle to an old person's hot water bottle – no one can replace me. Let's go back in time and discover my origins.

In ancient times, people would keep water in shells or leather bags. The first glass bottles were made in Egypt around 2000 BC. Molten glass was given the shape of a bottle by pouring it around a mould of sand. After the glass cooled down, the sand was taken out from it. For many centuries, bottles were made with this method only.

With the end of the Roman Empire, this art of bottle making lost its charm. In the Middle Ages, it reappeared in Venice, Italy, but not for any specific purpose, except for making showpieces. In the 17th century, manufacturing of glass bottles began in England in a big way. The glass could be melted in huge coal furnaces at much less cost. This made bottles much cheaper. Since these bottles were inexpensive and durable, they were used in storing water, wine, perfumes and every other liquid. England alone was making more than 3 million bottles every year – a record in itself. This craze reached the whole of Europe and slowly bottles became a household word.

In 1903, Michael J. Owens of Ohio brought about a revolution in the bottle industry. He manufactured a bottle-making machine, which could make 1800 bottles per hour as compared to the 25 bottles per hour made manually in the furnace. In other words, Owens' machine was equal to 72 workers! By 1909, Owens' company was one of the largest manufacturers of bottles in the world. Bottles were so popular in those days that once England's Prime Minister Winston Churchill suggested a war with glass bottles only.

In modern times, with the discovery of plastic, glass bottles have been replaced with plastic ones. Though they are lightweight and stronger than glass bottles, they are however not good for health. Environmentalists seek a ban on plastic bottles because plastic is a threat to the environment. A time may soon arrive when glass bottles once again become the king of the bottle industry.

14 Bricks

Hi! I am a brick. I am the building block of any building. I am strong, durable and dependable. Here goes my story...

Bricks are the symbol of advanced civilisation. So far many civilisations have been discovered. If bricks are found in its remains then that civilisation is said to be a well-developed civilisation.

Bricks have been in the making for more than 10,000 years. The Great Wall of China is the best example. This wall has been built with four billion strong bricks. In those days, clay bricks were dried in the sun. Later they were dried in kilns. The oldest brick was made in a kiln at Kalibangan village of India. Today this brick is kept in the Ancient Bricks Museum of America.

Bricks are the basic unit of a building. In ancient times, since means of transport were not well developed, bricks were built near the site of the building. When England decided to establish its own township in America, English sailors took three ships full of expert masons from England to make the bricks.

The process of preparing a brick is still the same. The clay is sieved, mixed with water and turned into a thick paste. Later it is given the shape of a brick and dried in the kiln. Earlier everything was done manually, but today everything is mechanised. The colour of the brick depends on the temperature of the kiln. Bricks containing some amount of iron also turn red in colour.

Collecting old and new bricks is an unusual hobby. But it is quite popular in foreign countries. Bricks belonging to ancient times or of various shapes and sizes are considered rare and sold for thousands of dollars.

Bricks are strong enough to have endured many revolutions in their lifespan of 10,000 years. But they have neither changed their style nor appearance, or even the technology. They have remained as solid as a rock and are still going strong.

15 Buttons

Can you imagine a shirt without buttons? Although a button very small, it has a special place in your life. Here's how is its story goes...

Buttons have been in use since 300 BC. At that time, buttons were used without a hole. They were fixed with ribbon on dresses. Till the 13th century, buttons were used in this manner. Only royal families or rich people used to have buttons on their dresses. Gradually, buttons came down to the common man and soon, everyone started wearing buttons on their dresses.

Even then, the size and style of buttons differentiated the rich and poor. While the rich wore gold and diamond-studded buttons, poor people had to make-do with buttons made of wood or bone. King Francis I of France had a special dress made with 13,600 gold buttons!

Later, gold, silver, copper, brass, ivory, stones and wood were used to make buttons. Intricate carvings made them all the more beautiful. Among all these buttons, brass buttons became very popular. They had pictures of the king or queen and were used on soldiers' uniforms.

Then came the Industrial Revolution and, thereafter, buttons were made with machines on a big scale. Matthew Boulton of England was the first person to make buttons with machines. Initially the response was not very good, as these buttons were expensive. Later, however, they became popular and their cost also came down drastically.

The boom in science and technology completely changed the process of button making. With the invention of automatic machines, thousand of buttons could be made in a single day. Celluloid and plastic buttons were now made with these machines, which were not only cheap but durable too.

Today, button making is not just a craft but a full-fledged industry.

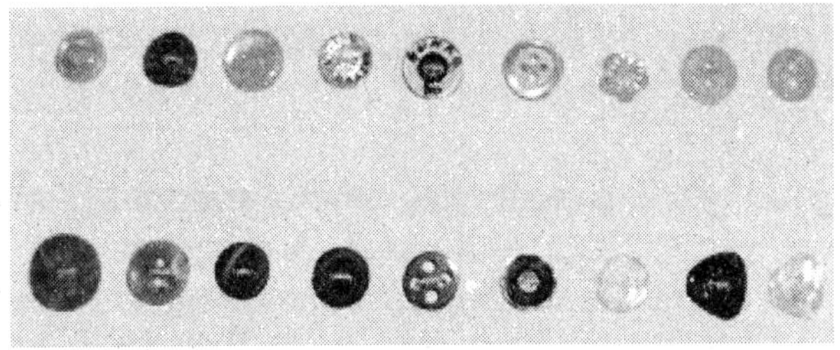

16 Chair

Hi! I give you comfort. When you are tired, you sit on me and relax. So, keep sitting till I finish my tale of origin...

Barring our face, a chair has almost everything a human body has. Hands, feet, back, seat. A chair has been in use since times immemorial. There is no evidence as to when and how the chair came into existence. All that we know is the Egyptians built the best chairs in the world. The golden throne found in the tomb of Tutenkhamen in Egypt is built on the basic design of a chair. Folding stools were also found in ancient Egypt. They were used by the rich while travelling.

Chairs changed their shape with time. In the Middle Ages, heavy chairs with intricate carving were popular. Today simple, lightweight chairs are in vogue. Folding and revolving chairs are also a gift of modern technique.

Earlier, kings used golden chairs studded with precious gems. Today P.M.s, presidents and premiers of big nations use revolving chairs. However, the truth remains that the chair was a symbol of power in those days as well as today.

Let's move on to the elder brother of the chair – the sofa. The word *sofa* has evolved from an Arabian word *soffah*. The siblings of the chair are the *couch*, (derived from a French word *coucher)* and the *divan*, (a Turkish word meaning *rest*.)

An easy chair is also meant for relaxation. It serves a dual purpose of a chair and a couch. Finally, there's the wheelchair, which benefits handicapped people and is a modified form of a simple chair.

17 Chemist

You must be familiar with today's air-conditioned chemist store where you can buy medicines. But this was not so till about a century ago.

'A dark, smelly dingy room full of huge glass jars with names written in Latin. Some more glass bottles with body parts of different animals preserved, a weighing scale, and some other bottles with red and yellow coloured liquid...' this is what a chemist shop in the 19th century looked like. This is not all. We need to go further down in history, if we want to discover the origins of the chemist store.

The world's first chemist shop was opened in Baghdad in AD 950. It sold herbs as medicines. People made fun of it because during those days most doctors treated patients with occult practices. So the shop was not very successful.

Later in the 13th century, a proper chemist shop was opened in Europe, which sold medicines on prescriptions. Since there were no patent medicines in those days, doctors wrote the method of making a specific medicine and the chemist would prepare it in his shop!

In India too, Ayurvedic doctors worked on a similar system, except for the fact that the doctor himself prepared the medicine.

Medicines prepared by chemists were substandard. So doctors decided to seek patents for their medicines. This was the beginning of modern medicines, which are packed and can be bought across the counter. In the absence of any control, these medicines had high amounts of alcohol. Finally, to curb these problems, the first law concerning patent medicines was passed in America in 1848.

Many reputed scientists were also attached to the chemist shop. Indeed, Sir Isaac Newton worked at a chemist shop. Similarly, American scientist Benjamin Franklin too sold patent medicines in Philadelphia.

18 Cheque

Have you ever wondered about the origins of the cheque? A cheque is a bank document that ensures you get the amount written on it from the bank. Cheques have been in use since the time when banks were not in existence. They were then called Bills of Exchange. Bills of Exchange were common in the 13th and 14th centuries and were mainly used by traders. Since most trade was by sea and to carry so much cash was a risky affair, people carried Bills of Exchange instead.

Some merchants took advantage of this system and started giving cash in exchange for these Bills of Exchange. They would earn a small amount as service charge for providing this service to traders. These merchants would sit on a bench and operate their business from there. So their business was later named as Bank.

By the end of the 17th century, the network of banks had spread all over the world. Slowly, banks started extending more facilities to customers like giving loans, taking deposits at a fixed interest rate and allowing cheque facilities. Earlier, cheques were not printed but were written on any piece of paper, even on the back of postal stamps.

Later, with the expansion of business, banks started giving out chequebooks as a safety measure. Even today, banks follow the same system. With the advance in technology, banks have been computerised and cheques also have an MICR (magnetic ink character recognition) code written on them, which is recognised worldwide.

19 Chewing Gum

Hi! I am a chewing gum. I am an all-time favourite of all people. Try me, you'll love the flavour. Here's how I was born...

It was in 1866 when General Santa Anna (the dictator of Mexico) was on the run from his country after a civil war. He picked up his belongings hurriedly and fled. But even in such a catastrophe, he remembered to keep a thick, white piece of gum made from the thick white juice of a Mexican tree. It was a tradition in Mexico to chew this gum during adverse times. Santa Anna sought shelter at Staten Island in New York.

A few months later when he returned to his motherland, that white piece of gum was left behind in his drawer. A scientist Thomas Adams, who lived on the same island, found this piece. When he heard that Anna chewed this gum, he could not believe it. He failed to understand what the stretchable thing was all about! He tried to prepare a new kind of rubber after vulcanisation but did not succeed. He tried to use it in sticking artificial teeth – again to no avail. Finally, Adams boiled it, placed its small pieces on the tip of tiny sticks and sold them to confectionaries as candy! The product was a big hit.

Inspired by this success, Adams realised he had hit the jackpot. Immediately, he built a machine that could make thousands of such gum-sticks and called his product *chewing gum*.

Before long, other merchants also entered this trade. Different colours and flavours were tried. Attractive wrappers were designed. Soon the market was flooded with chewing gums of all shapes, sizes, colours and flavours. To overcome this, Charles Flint built a chewing gum Trust. Six companies together formed this Trust. They captured the market in a big way.

But one merchant William Wrigley refused to join the Trust. He sold his chewing gums under the brand name Wrigley. He spent lavishly on advertising. At one point of time, he held 70 per cent of the total chewing gum market.

Today chewing gum is popular all over the world. A survey shows that each American chews 100 chewing gums a year. During the Second World War soldiers chewed 3000 chewing gums in a year. The proof of the popularity of the chewing gum is that there is not a single person in the civilised world who has not tried chewing gum at least once in his lifetime.

20 Chocolate

Hi! I'm your favourite chocolate. I bet you cannot resist me! But first you have to listen to the story of my birth. Then you can enjoy my delicious, sweet chocolate bar.

In 1519, when Hernan Cortes of Spain went to Aztec, Mexico, he was offered a sweet, scented drink in a golden goblet. This special drink was offered to honoured guests only. The natives called it *chocolate*. It was made from the beans of a tree. These beans were considered most precious in Mexico. The tribals believed it was a gift from God.

Though the story is probably apocryphal, there is no doubt that chocolate was a drink worthy of the gods. Its taste and flavour were out of this world. No wonder Carl Linnaeus named this tree as *Theobroma Cocoa* (meaning *God's food*). Today it is famous worldwide by its nickname *coco*.

Now back to Mexico... Cortes took some seeds of the coco plant and planted them in Spain. For about a hundred years, Spain guarded the secret of the coco plant. But with more and more explorers travelling around the world, the word got out and by the 17th century, entire Europe started drinking chocolate. Today Ghana, Brazil and Nigeria are the largest producers of coco.

Strangely though coco itself is bitter! Milk and sugar are added to coco beans to prepare sweet chocolate. Today, chocolate has captured the world market in a big way. Chocolates of different flavours, designs and sizes are available everywhere.

21 Cigar

We remember Christopher Columbus because he discovered America. But how many of us are aware that he was the one who brought the cigar to the civilised world? In October 1492, when Columbus' ship touched land, he was amazed to discover that tribals in the area smoked tobacco filled in big leaves. They called it *sik'ar*. Rodrigo Jerez, a friend of Columbus, tried this *sik'ar* and liked it immensely. Further inquiries revealed that the tribals had been smoking in this way for the past 500 years. Columbus brought these leaves back with him and called them *cigar*.

Spain was the first among European countries where the cigar became popular. When the army of England and France went to Spain with Napoleon they tried the cigar and took it back to their hometowns. Slowly, the cigar spread its wings all over Europe.

By the 18th century, America also joined the bandwagon. Bv and by, many cigar companies were established in America, experimenting with the design, style and size of the cigar. The world's smallest cigar was 1.25 inches long while the biggest was 6 feet 2 inches.

Cigars are famous for their special flavour and scent. Havana Cigars are world famous for their special flavour. They are made in Vuelta Abajo town of Cuba. It is said that their flavour is due to the climatic conditions of this area.

To smoke a cigar, there are certain rules too. One rule is that the smoker should not light the cigar with a lighter.

While lighting a cigar with a matchstick, the cigar should not touch its flame.

Winston Churchill was a great fan of Havana Cigars. During the Second World War, his huge stock of Havana Cigars was destroyed. When the shopkeeper he dealt with, heard of this, he called up Churchill at 2 o'clock in the night assuring him that his stock of Havana Cigars was safe.

22 Cigarette

Cigarette smoking is injurious to health. We see this statutory warning on every pack of cigarettes. Despite this, cigarettes have been with us since their discovery in 1518 in Mexico. Those cigarettes then consisted of hollow bamboo pipes filled with tobacco.

During the Crimea War in 1853, the English Army attacked a Russian train. The Russians on that train were carrying many homemade cigarettes. The English took away those cigarettes and brought them back to England. These cigarettes became so popular in England that many tobacco merchants took advantage of this opportunity and started manufacturing cigarettes. Soon New York and other European cities also got into the cigarette trade.

In 1890, James Albert Bonsak of Virginia built a machine that could fill tobacco in small neatly cut folded papers. Later these long paper tubes were cut to the size of a cigarette. These new cigarettes were convenient and affordable too. By now, the whole of Europe had become accustomed to smoking cigarettes. But people's health started suffering due to smoking. They became prone to chronic cough, which was termed *smoker's cough* by the doctors. This brought down the sale of cigarettes for some time but soon filter cigarettes were discovered, which seemed to be the solution to smoker's cough.

The first filtered cigarettes were made by P. Lorillard Company in 1952. Companies manufacturing filter cigarettes slowly controlled more than half the cigarette industry.

Despite claims to the contrary, the fact remains that cigarettes are harmful to health. They can cause throat, lung and mouth cancer.

Why, then, do people smoke? Scientific studies show that the nicotine present in cigarettes causes addiction, which a smoker is unable to shake off. But smokers find it very difficult to give up smoking. As the great humorist Mark Twain said, "There is no easier task than giving up smoking. I know this because I have given up smoking many times!"

However, it is up to you to decide whether you want to suffocate your lungs with the deadly smoke of cigarettes or give them a fresh breath of life.

23 Clock

Hickory dickory dock,
The mouse ran up the clock,
The clock struck one,
The mouse came down,
Hickory dickory dock.

Who hasn't heard this famous nursery rhyme based on the clock? Peter Henlein of Germany built the world's first clock in the 15th century. Henlein was an ironsmith by profession. It took him two years to build this clock, which was the size of a drum. It was six feet high and made of iron. This clock had just one hand, which indicated the hour. However, the clock was not too accurate. It was always an hour ahead or behind real time! These clocks hit the market in 1504. One of its specimens is still kept in the Philadelphia Memorial Hall Museum.

Size was a major hurdle as the big clocks were very inconvenient, and could not be carried easily. Many experiments later, a small watch was built which could be used as a pendulum in a chain. They were built in different designs and were quite popular at that time. Scotland's Queen Mary had one such pendulum watch shaped like a man's skull. The practical problem that made pendulum watches inconvenient was that it was difficult to see the time in these watches.

This led to the world's first wristwatch, which was made in 1581 for Queen Elizabeth I. It had a gold band studded with diamonds. But this watch also had a single hand for the hour. This technical problem was solved in 1676. For a long time, wristwatches were popular as a ladies' watch. Men generally used the pocket watch. During the First World War when soldiers faced a difficulty in using pocket watches, since their hands were filled with arms and ammunitions, the German Navy ordered wristwatches for them. Only their style and size were different from the delicate ladies' wristwatches. This ended the taboo and men and women both started wearing wristwatches.

Swiss watches are considered to be the best in the world. The world's first automatic watch was also built in Switzerland. Today's watches have three hands – for the hour, minutes and seconds. Alarm clocks led to the alarm in wristwatches too.

The modern watch is one with more gadgets in less space. One such model had a micro-speaker fitted in the watch, which would tell the time in a soft voice as if someone were speaking. Hey, now your watch has started talking to you!

24 Coffee

I have a special flavour. I may be bitter, but I am loved by people all over the world. So, tell me, isn't coffee your favourite drink?

Long ago in AD 850, in Ethiopia, a shepherd named Kaldi took his cattle into the fields for grazing. Here, some of his goats ate a few wild red berries. Soon thereafter, they started jumping up and down excitedly. Kaldi assumed that the berries contained some intoxicating product.

So he took some of those berries to the village priest. The priest boiled them and drank the extract. He found himself very fresh and full of energy. The priest continued drinking this juice even for his sermons during Sunday prayers. He noticed that while he felt sleepy during long sermons earlier, after drinking this refreshing drink, he now felt active and wide-awake. The priest shared his experience with other villagers and advised them to drink this juice. This is how coffee was discovered.

Coffee moved from Ethiopia to the Arab countries. In the 13th century, coffee was a popular drink in the Gulf countries. They even exported coffee beans. But they never sold the coffee plant or its seed for fear of letting go of their secret and losing out on exports. But secretly, some coffee plants reached different countries. Slowly, coffee plantation spread all over the world.

Coffee contains caffeine, which keeps the mind alert. This is the main reason why intellectuals prefer coffee to other drinks. Such coffee-addicted intellectuals built the world's first Coffee House in Europe.

Making coffee powder from coffee beans is a tedious process. Skill is needed for this and not everyone can do it. Then Instant Coffee powder was made, so that everyone could make coffee easily.

At one time, coffee was considered to be a medicine in France and England. From common cold to small pox, patients were advised to drink coffee. Coffee was so popular in Turkey that, at the altar, the bridegroom had to promise his bride to make the best coffee available for her till death did them apart.

Finally what is the definition of good coffee? Well, according to Charles Maurice Talleyrand, a good coffee should be *black like the devil, hot like hell, pure like an angel and sweet like love.*

25 Cups and Glasses

Hi! Wanna have a glass of milk or a cup of tea? If you do, then you need me. So have patience and listen to my story first...

In ancient times, man used hollow horns of animals, hard shells of vegetables, coconuts, eggshells and even a skull as a vessel. This went on till the third century when the art of glass making was developed in Rome. Now cups and glasses were made of glass. However, with the fall of the Roman Empire, this art was buried in the pages of history.

In the 16th century, Europeans revived the art of glass making. But there was a limitation. These cups and glasses could only be used in the local market because they broke when exported. To overcome this problem, large tin glasses with a handle were developed in America. These glasses were quite handy.

In the 17th century, glasses of animal hide were made. These were made by stitching the smoothened hide of animals and fixed with silver lace on all sides. Then came wooden glasses.

In India, disposable glasses and cups made of clay (*kulhads*) were also quite popular. The importance of disposable glasses was acknowledged during times of epidemic. In 1908, inexpensive paper cups and glasses flooded the market. These are still quite popular all over the world and only the paper has been replaced by plastic and thermacol. Today, all sorts of cups and glasses are available. You could pick and choose as per your need and your pocket.

26 Cupboards or Racks

I keep your things neat and tidy. You can call me cupboard, almirah or rack.

The world's first racks were built in caves when people felt the need to find a separate place to keep their belongings. Later on, as wood came into use, man built the world's first wooden rack.

With the passage of time, the size and shape of these racks kept changing. Then came a time when strong and intricately carved cupboards were in vogue. Even today, such cupboards are found in museums.

Today, light and strong cupboards are more popular. Such cupboards that are more spacious inside but take less space in the room are in demand today. There are even secret cupboards, which are not visible from the outside and are made to store valuables and important papers. Waterproof, fireproof, and earthquake proof cupboards are also available in the market nowadays.

27 Curd

Curd is an all-time favourite. It is nutritious and delicious too. Come let's see how curd came into our lives...

Man has been using curd since ancient times. It has been loved all over the world not only for its delicious flavour but for its medicinal properties too. Romans considered curd to be life-giving nectar. Curd was considered a cure for all ailments. Bald people believed curd could help them regrow hair. Old ladies would rub curd on their face to remove wrinkles.

All these myths prevailed till the 19th century before scientists began research on the properties of curd. Russian scientist Eli Metchnikoff discovered that curd contained some special bacteria. He called it *Lactobacillus Bulgaricus*. *Lacto* because curd was made from milk and *Bulgaricus* since the people of Bulgaria used to eat lot of curd.

Metchnikoff also noticed that the lifespan of an average Bulgarian was more than a hundred years. He concluded that curd was the secret of their longevity. So he too started consuming large amounts of curd each day. Unfortunately, he could not complete his centenary and died at the age of 71.

However, research continued to discover more about the wonder medicine curd. Soon, another bacteria named *Streptococcus Thermophilus* was discovered in curd. It was also learnt that these two bacteria were jointly responsible for turning milk into curd, making it more palatable and easily digestible. It is a scientifically proven fact today that curd is good for digestion and overall good health.

Curd is a good source of calcium, adding to its nutritive value. Today, curd is available in many flavours like chocolate, vanilla, kesar pista and so on. Gujarat's Srikhand and West Bengal's Mishti Doi are nothing but sweet flavoured curd, which are loved by everyone.

28 Cutlery

O dear, do you know I am nothing without my three best friends! Yes, the knife, fork and spoon are an integral part of any cutlery. Our history is very interesting. Would you like to hear us? OK, fork, you go first...

Forks

I was discovered in the beginning of the 17th century when an Englishman Tom Coryat went to Italy. There he saw people eating meat with a fork. He found it fascinating. When he came back to England, he presented a fork to Queen Elizabeth as a special gift. The queen liked the fork so much that she had its replica made in gold with jewels studded on it. When people saw the queen using a fork, they too started using a fork. This became a fad.

However, not everyone accepted the fork happily. Some were against it too, but the majority went with it. Slowly, it became a sign of prestige to eat with forks. People had special cases made to keep their forks.

During dinner invitations, sending your fork and spoon prior to the dinner was taken as a token of acceptance! This rule was broken when a Frenchman stated clearly in his invitation that guests need not bring their own forks, as they would be provided forks at the dining table.

Spoons

Hi, I am your friendly spoon! I am an old friend. Humans have been using me since the Stone Age.

It is assumed that after man discovered fire in the Stone Age, he started eating cooked meat. When the hot food burnt his hands he looked for alternatives, thus inventing the spoon. Spoons in those days were made with stone, bone or the shells of snails. Later, wooden spoons were made and these were in vogue for hundreds of years. They were improved with beautiful carvings. The word *spoon* has been derived from an English word *spon*, which means a piece of wood.

Later, after the discovery of metals, spoons were made from copper, brass or iron for the common man, whereas the rich ate with spoons made of gold and silver.

Knives

I may be sharp but my heart is as soft as jelly.

It is difficult to say when people started using knives for eating purposes, but it seems that the knife owes its origin to the fact that people would hunt with sharp knives and then later, they would pick their teeth with the same knives! This seemed very odd.

A Frenchman Cardinal Richelieu invented knives with a blunt end. Later, when the French Government banned sharp knives in the country, these blunt knives were used commonly. Still, such knives were not accepted in England till the end of the 19th century.

However, times changed and England also accepted knives as an integral part of cutlery. Today, the knife, fork and spoon rule dinner tables all over the world.

29 Diary

I am your secret keeper. I share all your joys and sorrows. Let's begin with the world's most famous and sensational diary...

Samuel Pepys lived in England. He was the son of a tailor but his talent took him to great heights and soon he became a Member of Parliament. He began writing his diary in 1660. He wrote religiously till 31 May 1669, when his eyes started deceiving him and he had to discontinue this practice. Many people write a diary, so what's so special about Pepys' diary? Pepys had written his diary in code language, which could only be understood by Pepys himself!

Even 122 years after his demise, his diary was still a secret. Then by a quirk of fate, someone managed to break the code of his diary and when the diary was read, it astonished the whole world. He wrote about his friends, colleagues and many other famous personalities who had formed his friendly circle. Everything was written openly with no holds barred and no secrets kept. The diary was such a testament of everyone's character that some parts of the diary could not be published.

It is believed that writing a diary is an art. It is a way of expression and it helps in removing stress too. Rousseau, Tolstoy, Emerson, Mahatma Gandhi and many other great leaders wrote diaries that have become a source of inspiration to the whole world today. Charles Darwin wrote his famous *Theory of Human Evolution* on the basis of findings noted in his diary. In 1831, when Darwin went on a sea voyage to South America on the HMS Beagle, he carried many diaries with him to note down important findings.

Another famous diary is *The Diary of Anne Frank*. Anne Frank was a Jew. She wrote about her horrible days during the Second World War when she had to spend two full years hidden in an attic for fear of being captured by Hitler's Nazis. How she died a hundred deaths each day has been described in a heart-rending account given by Anne Frank. After her death at the hands of the Nazis, the poignant account in her diary has survived to tell her sad tale about the horrendous atrocities of the Nazis.

30 Dice

Hi! Wanna play a game of Ludo? Then get a dice please!

Do you know what is a dice? From ancient civilisations right up to our modern civilisation, the dice has always been a part of our lives. Ancient civilisations had dice of difference sizes and shapes. Though most of them were in the shape of a cube, like the ones we use today, dice in square, rectangular, pentagon and octagon shapes were also found in those days. A dice in the shape of a pyramid has also been found.

In the excavations in Egypt, dice as old as 2000 BC have been found. Dice from 600 BC have been found in China. A book of religious hymns of Hindus, the *Rig Veda*, mentions dice. The great Indian epic *Mahabharata* was based on the game of dice. If there were no dice, there would have been no *Mahabharata* or *Bhagavad Gita*.

Some games are played with a single dice while others use a double dice. There are even games in which five dice are used. Children's favourite indoor games like Ludo, Snakes and Ladders, Monopoly and Trade are also played with dice. Even the arrival of the computer has not taken the fun out of a game of dice. Modern kids are equally crazy about these games as their ancestors once were.

31 Dictionary

Ask any school-going kid and he will recognise a dictionary instantly, such is the popularity a dictionary enjoys. Would you like to know how a dictionary came into existence?

A dictionary is a thick book that covers almost all words of the language. The French dictionary took 56 years to complete. The first authentic English dictionary, *The Oxford English Dictionary*, took 71 years. The German dictionary was 106 years in the making. An Italian dictionary begun in 1863 is still incomplete!

Amazingly, the originator of *The Oxford English Dictionary* was not a scholar. He had studied only up to Class VIII. The son of a simple village tailor in Scotland, James Augustus Henry Murray loved playing with words. He was a genius and could recognise many English words when he was just one and a half years old. He loved to study, but due to unavoidable family circumstances he could not attend school after Class VIII. But the entire village knew about his intelligence and he was appointed the Head Master of the village school by the time he was 20. It is said that he had learnt 20 languages by then and his hold on English was remarkable. All this made him famous.

Soon, the Philological Society of England contacted him and gave him the task of preparing an English-English Dictionary. This project had already started in 1857 but was lying untouched for the past 22 years. Murray's estimated time was ten years at the most. But when he checked the reference material, he was stunned. There were more than 2.5 million reference slips! Murray had to build a shed in his garden to keep these slips safely. He spent all his time in writing the dictionary. Murray's 11 children also helped him in this venture. Sometimes, a single word took several months! The word '*Do*' took six months.

In 1915, Murray died at the age of 78. The dictionary was not yet complete and was written only up to the letter T. The complete dictionary was only published in 1928. This *Oxford Dictionary* had 4,14,825 words and till that day £2,91,000 had already been spent on it.

Before the *Oxford English Dictionary*, the *Samuel Johnson Dictionary* had come into the market in 1733. But it was not very successful as it had many shortcomings. William Caxton wrote the world's first dual language dictionary in 1480. William Caxton is also famous for inventing the printing press. He prepared this dictionary for the convenience of tourists.

32 Dry-cleaning

It is so easy to get your expensive clothes dry-cleaned today. But it was not so until about 200 years ago. The year was 1825. In France, a maid was working in Jean Baptiste Jolly's house when suddenly a camphene (turpentine oil) lamp, kept on a table, tripped and the oil fell all over the tablecloth. The maid was scared and rubbed the tablecloth with her duster to remove all stains of the oil. Surprisingly, the more she rubbed the tablecloth, the cleaner it became. Soon the whole cloth was as good as new. She informed Jolly about this event.

Jolly was a wise man. He understood the importance of this oil and used this technique in cleaning expensive clothes. He opened a shop where he cleaned clothes with camphene. He called it dry-cleaning because there was no water used in the process.

Slowly the process of dry-cleaning became popular worldwide. England welcomed it with open hands because washing clothes was a very costly affair there. But there were two problems with camphene. First of all it was highly inflammable and, secondly, its smell would not leave clothes easily. People tried to replace camphene with naphtha, benzene and gasoline also but to no avail.

Later, scientists experimented with carbon tetrachloride but had to abandon its use since it was very poisonous. Finally, perchloro-ethylene was accepted worldwide as the best dry-cleaning chemical. Even today, clothes are dry-cleaned with perchloro-ethylene.

Do you know the uses of dry-cleaning? It not only cleans your expensive and delicate clothes but can also clean heavy drapes, leather jackets, blankets, carpets and animal furs too.

An elephant was once dry-cleaned in Massachusetts! This elephant was coloured shocking pink during an advertising campaign. Later, when its owner tried to wash it with soap and water, the colour would not go. Ultimately, the owner contacted the dry-cleaning shop as a last resort. And sure enough, the elephant was as good as new! Its skin was not harmed either. So that is the power of dry-cleaning.

33 Dye Colours

Colours make our lives beautiful. Do you know how dye colours entered our world? It happened in 1856 at London's Royal College of Chemistry where an 18-year-old boy William Henry Perkins was studying chemistry. Perkins was very inquisitive and performed various experiments in his spare time. At that time he had been working on a medicine for malaria.

One day while conducting an experiment, Perkins mixed some chemicals and a thick black lump was formed. To know what it was, he put it in alcohol. The alcohol became shining purple. When he put a silk cloth in this solution, it became purple too. The colour was permanent. Perkins sent this cloth to a famous dyeing company in England seeking their opinion. Their reply was quite positive – if ladies liked your colour and what you claim proves to be correct, then you are on your way to becoming a millionaire.

Luckily, Perkins's dye, which he had name Mauve, found favour with Queen Victoria. She had many of her clothes dyed in this colour. The whole of England turned mauve. This was the world's first artificial dye.

Natural dyes were in use since the Middle Ages when colours were extracted from trees, plants, and flowers and even from tiny insects. Natural purple colour was extracted from the mucous of a special snail, which was very expensive. Yellow and orange colours were found in many plants and flowers. Deep red colour was obtained from a tiny insect called Dactylopius Coccus but was very expensive because one pound of this colour was obtained after sacrificing almost a million such insects.

Deep blue colour was obtained from the woad plant. But it was not very popular because of its stinking odour. It was so smelly that the Queen of England had banned the extraction of blue colour from woad for five miles around her palace.

Artificial dyes were a blessing in disguise. They were neither smelly, nor expensive. Soon they captured the market. Dyes of different colours were also developed in the course of time and people could now enjoy the colours of their choice.

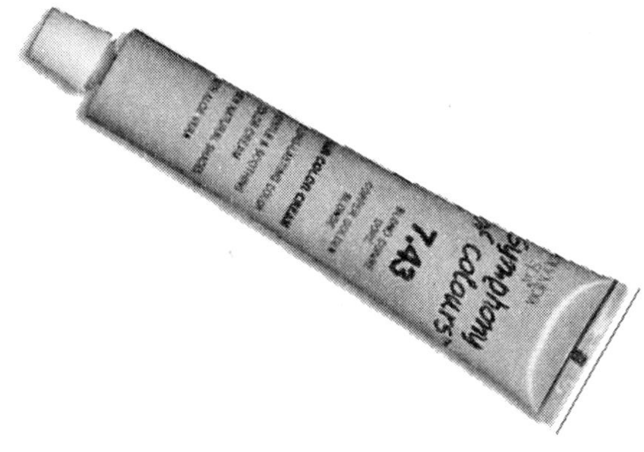

34 Electric Bulb

Hi! You are reading about the story of everyday things. But do you realise that if I were not there, many everyday things would not be possible? So check out my origin in the following lines...

Thomas Alva Edison was a genius who started making new useful things at the age of 16. He invented the electric bulb, which he found rather simpler to make. He made this bulb in 10 days.

But what took more time was how to place an element inside the bulb. No filament could take the high load of electricity. He tried all materials available to build it. He even tried a hair from his assistant's beard, but to no avail. The 10 days then stretched to 14 months.

Finally he managed to make the world's first bulb by using an ultra-thin thread of cotton. It worked for 40 hours. Soon Edison was famous and his name sparkled with the illuminating electric bulb.

Today elements are made of tungsten. It can bear the heat of 6170 Fahrenheit. Thinner than a hair, this element is no bigger than half an inch. Now, sodium vapour and mercury vapour bulbs are also made, which are much more powerful and durable than ordinary bulbs.

Ultraviolet bulbs are also made, which help in finding many hidden minerals, insects and water from inside the earth. It also helps in detecting fake paintings. These bulbs are also used for giving heat to plants in the absence of sunlight. It is said that such crops give better yield. Even poultry flourished wherever these ultraviolet bulbs were used.

35 Envelope

Fill me up,
Stick me up,
And hand me down.
All over the town.

In days bygone, as soon as man learnt to read and write, he needed a proper cover in which he could send his letters safely. Earlier envelopes were made of similar material used for writing, like cloth, animal skin, tree bark etc.

In Babylon, the letter would be covered with a thin layer of clay and then heated on fire. The clay would become hard and act like an envelope! Later, when paper was invented, envelopes were made of paper.

In 1653, D. Valayer of France started a postal service in Paris with the permission of King Louis XIV. He placed letterboxes at each crossing. He sold his own envelopes and letters sealed in those envelopes could only be put in those boxes. But this service could not run for long, as the letters taken out of the letterbox were found to be torn. People lost faith in him. Later, it was revealed that Valayer's enemies would put a rat in the letterbox to destroy all the letters. So this animosity led to Valayer closing down the world's first postal service.

In the beginning of the 19th century, England started its own postal service. At that time, the receiver had to pay the postal charges. Some clever people devised dubious ways to save money. They created many codes, which would be written outside the envelope. When the postman brought that envelope to them, they would read the code written outside the envelope, get the message, refuse to pay the money and send the letter back. The creation of postal stamps put a stop to this practice.

Change in technology has altered the style of envelopes too. Now self-adhesive envelopes are available in the market. There are water and chemical proof envelopes too. The paper used to make these envelopes is much lighter and stronger. It is impossible to tear these envelopes.

36 Flag

A flag is a symbol of of freedom and integrity of a country. The Indian flag is also called the Tricolour.

Flags have been in use for a long time. However, the world's first flag did not belong to any country but was used to ward off evil spirits. Such flags had a special sign marked on them, as directed by the village priest. People would assemble under these flags and feel safe from evil spirits!

Later, these flags became the symbol of a particular village or tribe. As the tribes combined, city-states were formed and, gradually, countries came about and national flags then came into existence. By the 10th century, most countries had national flags. In battlefields also, the king and commander's presence were marked by their flags. In the absence of these two, their flags were protected by the army. If the king's flag fell, it was a clear-cut indication of defeat.

Each country's national flag denotes something. There is an interesting story behind Austria's red-coloured national flag with a white stripe in it. It is said that in a fierce battle, the army of Austria was defeated badly. A handful of soldiers were trying to protect their flag, when it fell down. Immediately, one of the soldiers took off his shirt and put it up as a flag. The white shirt was covered with red blood, except for the place where he had tied a band. Later, in honour of this soldier, Austria declared it their national flag.

A national flag is a country's symbol of honour. Insulting a national flag is a punishable offence.

37 Gates and Doors

There is a bronze gate at a temple in Rome which is about 1800 years old and as strong as new. Another gate in a church in Florence, Italy is also as strong with intricate carvings. Lorenzo Ghiberti, a great artist, carved this gate through hard labour of 27 years. Even Michelangelo was amazed at Lorenzo's beautiful work. He named those gates as the 'Gateway to Heaven'.

Doors became popular as soon as man felt the need for privacy. Even cavemen would hang the skin of animals or branches of trees on the mouth of their caves. If they went out for a longer period, they would cover the cave's opening with a huge stone. Later, when man began living in huts, he started using cloth curtains or mat doors on the main entrance. With strong houses, wooden doors became popular. Iron doors were also built but were mainly associated with prisons or high security areas.

In the Middle Ages, beautiful wooden doors with intricate carvings were made. The main gate of a house defined the status of the owner. The main gates of forts were made of iron and would be so heavy that many hands were needed to open them. These doors were fitted with sharp knives from outside to keep the enemy away.

Times changed, and heavy, strong, big doors went out of fashion. Today, lightweight aluminium doors are preferred to heavy wooden doors. Carving too went out of fashion as it is expensive and serves no purpose. Sliding doors were first made in Japan. Today they are popular all over the world as they save space. Folding doors are also used in places short of space.

Revolving doors were made in Philadelphia, USA, about 100 years ago. They were more popular in areas where extreme weather conditions prevailed and were popularised by saying that these doors efficiently kept snow, water, dust and breeze out.

Doors which open automatically after a certain period or ones which open automatically at the sound of approaching footsteps are the gift of modern technology. Now the ultimate step in the story of doors will be a replica of the 'Open Sesame' fame of Ali Baba and the 40 thieves fame, which will open with a voice command only. For once, it will be a dream come true.

38 Gramophone and Stereo

I am a magic box. I sing and make you dance. Guess my name... Yes, I am the Gramophone. Let me tell you my melodious story...

Almost 3000 years ago, a Chinese prince had a magic box. He would open the box, speak a message for his friend into it and close the box. Then he would send the box to his friend. When the friend would open the box, the prince's voice could be heard...

Of course, this is simply a famous Chinese fable. But in 1632, a captain who had returned from a sea voyage around the world told the people that he had seen the residents of South Sea Islands talking to sponges. When those sponges were pressed, the same voice was repeated. Though he could not prove it, still it shows how keen man has been to listen to the echo of his voice.

In 1877, Thomas Alva Edison built the world's first gramophone without the aid of any sponge or magical powers. The voice was recorded on a tin strip loaded on the drum of the gramophone. The first words recorded on the gramophone were '*Mary had a little lamb...*' sung by Edison himself.

However, the quality of the drum was substandard. Neither was the voice clear nor was the tin strip durable. Later, wax drums replaced the tin strip. These were further improved upon and now records were used for recording.

Edison was hard of hearing. So he did not much care for music. He reckoned gramophones could be an educational aid by recording text and other study matter. At that time, he could never have imagined that his invention would give rise to the multi-million-dollar music industry.

The stereo is an improved version of the gramophone and was developed in the 1960s. Recording in stereo mode is done using more than two microphones. Before the invention of magnetic tapes, recording was a tedious process. Even a minor mistake would be apparent and the whole song had to be recorded again because there was no way of rectification of errors.

Today, all these problems are a tale of the past. Recording is now smooth and hassle-free. Indeed, life is melodious and stress free, thanks to the gramophone and stereo.

39 Hamburger

Hi! I am a hamburger. People may call me fast food but they eat me all the same. Here's how I was born...

The hamburger sounds like a product of Hamburg, Germany. But you are wrong here, my dear readers. Tartar Nomads created the hamburger or burger. Nevertheless, it was named hamburger much later when the traders took it from Asia to Germany. And they named it Hamburg Steaks, which became hamburger and today it is more popular by its nickname burger.

During the Middle Ages, the Tartar Nomads would eat raw minced meat. Sometimes, when the meat was hard, they would make a circular ball of minced meat and keep it under a heavy metal so that it would soften up. After some time, they would take it out, mix it with salt, chilly and onion and eat it. Today burgers are found in vegetarian flavours also like cheeseburgers, veg burgers, bean burgers and aloo tikki burgers.

Hamburgers are most popular in America. More than 53 billion burgers are consumed in America itself every year. The world's largest hamburger was made in Australia in 1975 – it weighed 2,859 pounds and had a diameter of 27.5 feet – the mammoth hamburger, we must say.

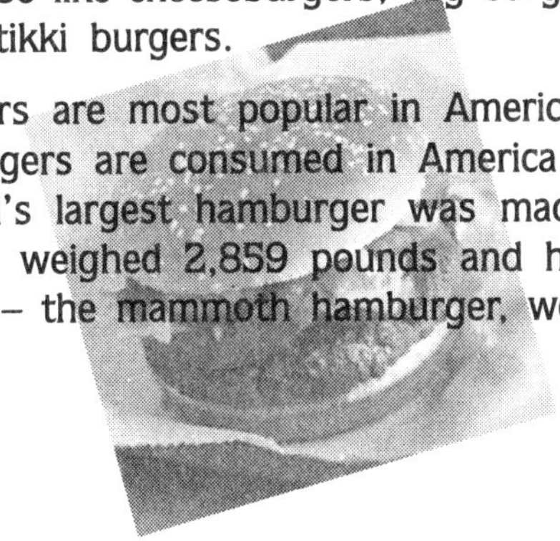

40 Ice Cream

Children love it, and adults relish it. Be it summer or winter, ice cream is an all-time favourite. Have you ever wondered how this delicious concoction came into existence? Let's discover how...

Alexander the Great loved eating ice cream. In those days, ice cream was made with honey, fruit juices and milk. Ice from the mountains was brought to make the ice cream. Hundreds of slaves used to work day and night to prepare a cup of ice cream for the great emperor.

The art of making ice cream was developed in China around 3000 years ago, from where it spread to India and the Gulf and Arab countries. In 1295, when Marco Polo went back to Italy after an excursion to China, he praised the flavours of ice cream to his countrymen. But Europe still had to pass some more time without ice cream.

In 1685, when the marriage of the princess of France, the daughter of King Henry IV was fixed with King Charles I of England, for the first time the Queen of France brought ice cream to England. Charles loved the ice cream. He offered a special reward to those workers who made the ice cream, so that this art could be kept secret. Charles wanted ice cream to be a royal dish. He did not want it to reach the common man. But this was not to be. Soon people got to know of this scrumptious dish and enjoyed it. However, ice cream was not easy to make.

In 1846, Nancy Johnson of America built a manual ice cream making machine. It was a double-walled box. She would fill ice between the two walls and keep the ice cream mix inside the box. In 1851, with the help of this instrument, Baltimore established the world's first ice cream factory and started making ice cream on a large scale. Soon another fact came to light – if you add salt to the ice, the temperature falls quickly and ice cream could be made faster.

The 20th century saw ice cream hit the market in a big way. It could either be served on a plate or in a cone (commonly known as softy). How the softy came into existence is another interesting story. In 1904, in a fair at St. Louis, an ice cream vendor suddenly found himself short of plates. To overcome the crisis, he borrowed thin wafers from his neighbouring shop, folded them in the shape of a cone and started selling them filled with ice cream. The public found it delicious. Soon it became a rage.

Ice cream on a stick was started in 1911 when a child wanted to buy an ice cream as well as candy but did not have sufficient money. However, he bought the ice cream in the end. But this gave the shopkeeper an idea. He experimented by selling ice cream on a stick, which became very popular.

So now you know how ice cream came into existence and how it changed its forms. Interestingly, today ice cream is sold in all forms because:

I scream, you scream
We all scream for ice cream.

41 Jeans

I am the strongest. To prove my point, let me tell you this real incident.

It happened years ago. A labourer was working on the 52nd floor of a building in Texas, when suddenly his pocket was stuck in a crane's hook. The crane picked him up. His death was merely a pocket away. It was clear that the moment his pocket tore, he would fall 52 floors to earth. Amazingly, his pocket did not tear and within minutes the crane had placed him back to safety. The strong pocket saved his life. He was wearing a Levi Strauss Jeans.

Subsequently, he thanked the Levi Company for saving his life. He was so impressed that he complimented them by saying that if a link between two bogies of a train broke, it could safely be repaired with Levi jeans.

Jeans were born in 1850. When young Levi Strauss came to San Francisco, he brought lots of canvas to build tents for labourers working in gold mines. He was doing fine until one day a worker jokingly requested him to make pants for him with that canvas because his pants got torn in the mine very quickly. Nonetheless, Levi stitched the pants for him and it was a great success.

Soon all the workers were wearing Levi's canvas pants. Levi abandoned his previous business of making tents and started making pants instead! When the canvas stock was over, he made denim pants. He coloured them blue. The workers liked these too. For the next 20 years, Levi's business flourished in this manner.

In 1872, a Russian tailor Jacob Davis wrote to Levi Strauss. He suggested brass rivets on the seams of these pants. This would make these pants stronger. Davis suggested a partnership. In 1873, they opened their own company by the name of Lewis Strauss and Company.

Till 1920, however, only workers wore these pants. Later, some film stars and other renowned people wore Levi jeans, and it became a fad. In 1940, some schools designed their uniform with jeans. Since then, jeans have been the favourite dress of the young generation.

As jeans reached higher circles, it was modified as per their status. Famous film star Lana turner had diamond-studded jeans. Henry Kissinger also wore the best-fitted jeans. The Levi Strauss Museum in San Francisco displays a pair of jeans worn by a woman regularly for 17 years. The jeans were completely worn out, yet its seams were intact.

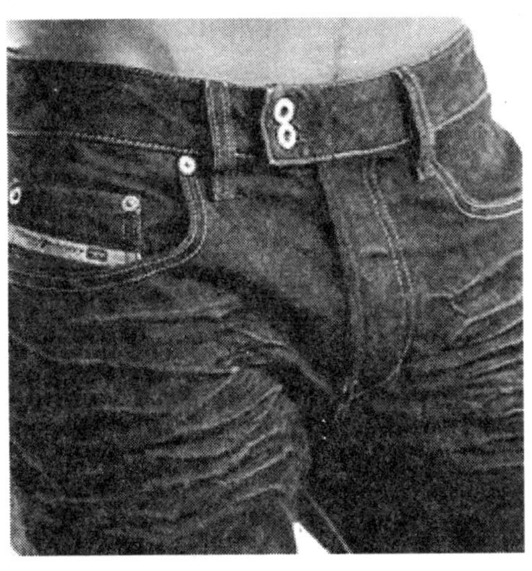

42 Jewellery

Hi! We are expensive, yet we enjoy a great rapport with all people. We are ornaments adorned by people all over the world. Our story begins like this...

Ornaments have always enchanted humans, whether made of stone or of gold and diamonds. The Stone Age had ornaments of stone, while today they are made of various substances like gold, silver, platinum, and diamonds. The main aim of wearing ornaments is to look beautiful. Garlands of flowers also enhance beauty. From time to time, women as well as men have been wearing contemporary ornaments to beautify themselves.

In Babylon, some stones were worn to keep away evil spirits. Soon, these were categorised into birthstones, according to the 12 months of the year. The Romans believed that the salt of precious stones could ward off many ailments. Harder stones that could not be ground were worn around the neck. This was the world's first necklace.

People travelling to distant lands were given a Royal Seal, which was akin to today's passport and visa. Afraid of losing the precious seal, people started wearing it around their fingers. This was the world's first ring.

Pearls hold a special place in the world of precious stones. They were believed to be the frozen dewdrops or tears of a fairy. Pearls were considered very precious. The world's first pearl jewellery was made in Rome.

Each big diamond in the world has a story attached to it. Who is not aware of India's world famous diamond "Kohinoor"? It has been stolen many times from our country. Finally, it now rests in the crown of the Queen of England.

Another such saga surrounds the Hope diamond of France. It belonged to King Louis XIV. It is believed that whoever possesses this deadly diamond does not live long. Either he dies in an accident, gets killed or commits suicide. Today, it is kept in the Smithsonian Institute of America.

Expensive ornaments worth millions but which are so small that they can be safely tucked away in a tiny locker, have always fascinated thieves. So beware while wearing real ornaments.

Replicas of real ornaments, or imitation jewellery, are also quite common today. They not only look equally good but are easy on the pocket too. There is no fear of theft or burglary either. However, nothing compares with the real thing.

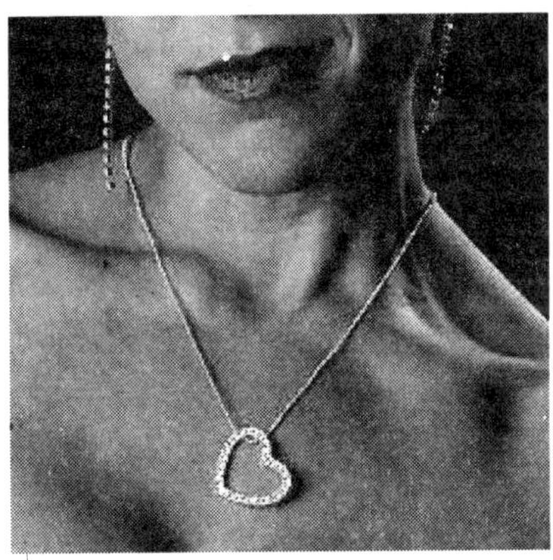

43 Lifts

Hi! I am the majestic lift. When first invented, people would pull me with a hand-driven winch, because till then they were not aware of the electric motor. I owe my modern and safer looks to Mr Otis – a friend, philosopher and guide. He made me what I am today. So here is my story...

The world's first lift was described by Roman engineer Vitruvius in 26 BC. He called it the flying chair. Much later, in 1743, a similar flying chair was installed outside the chamber of French Emperor Louise XV. Riding on this chair, the Emperor would reach his Queen's chamber on the upper floor within seconds. This was considered to be the first lift in the world.

In 1854, a brave American E.G. Otis performed an amazing feat when he boarded an open lift in a high-rise building of New York. As the lift started moving, he cut the rope of the lift. The spectators stood there mesmerised, waiting for the lift to fall to the ground with a bang. But... nothing happened. After some time, the lift stopped on its own. No unpleasant incident took place and Elisa Graves Otis became a hero.

In fact, this was a risk Otis took to promote the lifts manufactured by his company. He proved to the world that even if the rope breaks, lifts are not risky. This event was popularised in a big way. Yet, for three years, Otis could not sell a single lift. In 1857, a five-storeyed departmental store in New York got the first lift installed. The lift moved with steam but it was very slow. In one minute's time it moved 40 feet only, so this lift was not feasible for high-rise buildings.

To increase the speed of the lift, many experiments were done. However, this problem was only sorted out once lifts were electrically operated. The first electric lift was installed in Demarest Building of New York in 1889. Since then, there has been no looking back. Now lifts are so advanced, they can move up and down by the mere pressing of a button. They are also automatic. It is also easier to call the lift on any floor by just pressing a button. And the best part is that the Otis Company has proved its monopoly by making all these improvements in lifts on its own.

However, lifts have become more expensive over the years, considering the first lift of Otis was sold for $300, compared to their modern fully equipped lifts sold for $150,000. But today what matters are the quality, speed and other modern gadgets that are fitted in lifts. Lifts are a necessity in today's urban lifestyle rather than a luxury. One cannot imagine a high-rise building without a lift. Such is the power of a lift.

44 Lock and Key

They're true friends. They work together. That's why there is a key to every lock.

The story of locks is about 4000 years old. The Khorsabad Fort of Iraq boasts of the world's oldest locks. This fort was built in 2000 BC and around this time, Egypt had also invented a lock system.

The old locks were huge and made of wood. Their keys were in the shape of a toothbrush. Such locks were found in Japan, Scotland, Norway, Egypt and India. The wooden locks were first made in Rome. They were fitted inside the door with just a hole for the key showing from outside. The ruins of Herculaneum and Pompeii cities have remains of such locks. The keys to some of these locks were so small that they could be worn like a ring around the finger.

More locks were invented later, but were not very reliable as their duplicate key could be made very easily. Locks continued their journey and kept improving in the process. By the turn of the 19th century they were much smaller and easier to operate.

But a shocking thing happened around 1850 in America. Americans were stunned to find all their Parautopic locks (famous locks of that time) being opened by an ordinary young man easily. He neither tampered with the lock, nor broke them; he simply opened them with a master key, which fitted all the locks perfectly! Sometimes, he would press these locks so tightly with his master key that even the original key could not open the lock! The company faced a severe setback.

This man, Linus Yale Jr, was not a conman. He was a painter who had inherited this knowledge of locks from his father Yale Sr, who was an expert in locks. However, this feat won Yale Jr the title of 'magician of locks'. In 1861, Yale Jr made a special lock called the 'cylinder lock'. These locks were special because no one lock could be opened with another lock's key. Our modern locks are an improved version of these locks.

Yale Jr also invented 'number locks', which were a great success. The only lacuna in the number lock was that if someone knew the number, he could easily open the lock. To overcome this problem, Yale Jr came out with the 'time lock' system.

The world of locks is filled with amazing locking devices. There are locks that can sense their owner's voice in a similar way like the 'open sesame' of Ali Baba and the 40 thieves. It is amazing how locks control our lives and our treasures without our realising this.

45 Lottery

Hi! Wanna get rich quick? Then buy a lottery ticket and fulfil your heart's desire.

'Who shall fight Hector, the Prince of Troy?' The question loomed large over the soldiers. Finally, a lottery was drawn. All the soldiers put down their names on small paper chits and placed them in a hat. One chit was selected and that person was sent to fight Hector."

This is an extract from Homer's great epic *Iliad*. Even the Bible's *Old Testament* describes the lottery system while distributing land. History is full of such events where a lottery was drawn to decide something. No one knows the origins of the lottery, but it has been around for a long time.

Earlier, the lottery was drawn without involving money, till one fine day when the great philosopher Augustus began a new rule, which stated that all those who wanted to be party to the lottery system had to pay some amount to the treasury to get their names enrolled for the draw. Till date this system is prevalent.

In the Middle Ages, a lottery system was developed under which the amount collected after selling the lottery tickets were divided into two parts. One part was for the winner and the other part was used for the welfare of society.

King Francis I of France introduced the first big lottery in 1539. Queen Elizabeth I followed suit and, in 1566, England also declared its first lottery. There were 40,000 tickets to each lottery draw costing 10 shillings each. These lotteries gained popularity since they were promoted for charity. Though a lottery is a gamble, but the clean and charitable motives advertised by the brokers freed participants from any guilt of gambling. And it showed too. The world famous British Museum is built with money earned through lottery. Harvard, Columbia, Yale and Princeton Universities have also been built with lottery money only.

The lottery started as a hobby but in course of time it became an obsession. Charitable motives were also buried under man's growing desire to earn wealth through fair or foul means.

Let me narrate a true story about lotteries. A Frenchman was obsessed with lottery tickets. He bought thousands of tickets during his lifetime but never won any bumper prize. When he died, he was buried in his best suit. When his widow was screening his papers, she came across a slip of paper with a lottery ticket number scrawled on it. Out of curiosity, she enquired about this lottery, whose draw opened the same day. Voila! She had hit the jackpot. The ticket had won first prize. But where was the ticket? She searched the whole house but could not find the ticket anywhere. Now the only place where that ticket could be was her husband's grave. It struck her that the ticket could be in his suit pocket. So the fresh grave was dug up again and she found the ticket in the coat pocket!

Indeed, sometimes dreams do come true. But some dreams come true only after death. Too late!

46 Maps

Hi! I tell you the directions and help you reach your destination. Here is how my story goes...

All the maps made before the 17th century were incomplete because they could indicate the directions but not the distance. After the invention of various aids and instruments like the sextant, proper maps were made.

The world's first and the biggest map is the globe. Before the globe or paper maps were made, explorers would draw their own maps on cloth and guide people with these cloth maps and show them the places they had visited during their voyage. Since they were made on cloth, they came to be called *map* because *meppa* in Latin means *napkin*. However, some old maps made on animal skin have also been found.

Famous Spanish explorers Columbus and Magellan had also charted various maps to reach the new world. It was believed that this new world was very rich and full of gold and silver. So these maps were stolen from a museum in Spain.

Maps of different sizes and for each and every part of the world have been drawn today. There are maps to find hidden treasures too. Secret signs are used to indicate the places near hidden treasure. Even the armed forces use various maps for their strategies and attacks.

Many adventurous films have also been made based on solving the complicated maps. Till date, there have been many maps found throughout the world that are yet to be deciphered.

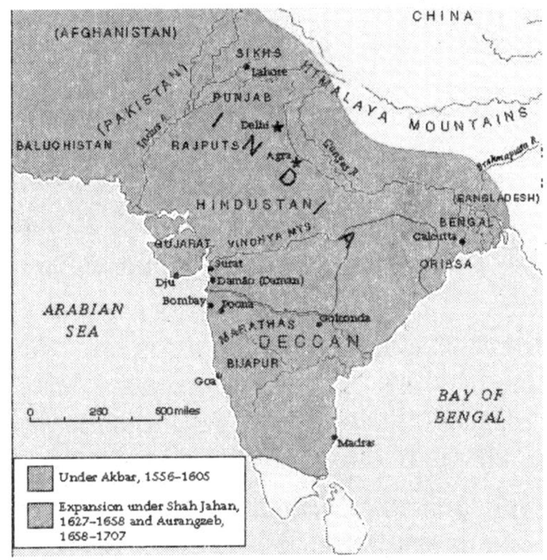

47 Matchbox

Hi! I am the simple matchbox. I am full of matchsticks. But beware – if you play with me, I can burn your fingers.

Henning Brandt discovered phosphorus in 1669. It was used in fireworks at that time because of its highly combustible nature. Since it was very expensive, only rich people could afford to enjoy such fireworks.

In 1827, John Walker of England made the world's first matchbox using phosphorus. He called it Lucifer, which in Latin means *to light up*. Lucifer had to be rubbed against an abrasive paper, which produced a spark and lot of smoke that smelt like a rotten egg.

However, it was soon discovered that phosphorus was harmful for health. Those working in matchbox factories were the worst sufferers. Some suffered disfigurement of bones, while others suffered fossy jaw. Some even died. Labour became an acute problem in the matchbox industry. In 1911, the Diamond Match Company solved this problem when they started using sesqui sulphide in place of white phosphorus. This was as good as phosphorus and had no harmful side effects. A new era began in the world of matchboxes.

It would be unfair not to mention Ivar Krueger's name when we talk of matchboxes because he controlled about three-fourth of the world's total matchbox market in the 1900s. Actually he was a crook. Soon the world saw his real face and he was imprisoned. Krueger committed suicide at the age of 52 when he could not find any other way out.

48 Mirror

Mirror, mirror on the wall. Who is the fairest one of all?

Mirrors always tell the truth. When man saw his reflection in the water for the first time, he was frightened. He thought some wicked demon was hiding in the water. This led to the belief that demons lived in mirrors. In ancient times, people in China would hang copper-plated mirrors on their doors to ward off evil spirits. In Japan, the culprits were made to confess in a mirror. The judges would assess a culprit by studying his expressions in the mirror.

All these misconceptions were cleared when Euclid, a great mathematician, discovered the Optical Laws. He stated that mirrors act according to these laws. Later, the great scientist Archimedes invented concave mirrors. Initially, he used them in war. When Rome attacked Syracuse, Archimedes placed huge concave mirrors at such an angle that sunrays falling on these mirrors burnt the Roman warships.

Many centuries later, the great scientist Isaac Newton invented the world's first reflecting telescope with the help of concave mirrors. These telescopes are so useful that today astronomers can study distant celestial bodies from thousands of kilometres away.

49 Motor Car

Wanna ride me? When the Stanley brothers invented me, no one wanted to ride me but today... Well, it's a different story altogether.

F.E. Stanley and F.O. Stanley were twin brothers in America who built the world's first steam-operated car. One fine day, the car ran at a speed of 197 miles per hour and then rose high up in the air. This flight stretched for about 100 feet before the car collapsed back to the ground. Miraculously, the driver survived this ride-cum-flight. Since that day no one dared try to ride this car. Not even the Stanley family that had proudly used their car till now. The car was locked in the garage.

The first modern gasoline engine was built around 1870 by Nikolaus Otto of Germany. With the passage of time, its design was modified and by the end of the 19th century, the roads of England and America were flooded with motorcars. Strange traffic rules for the motorcars were made that could make one laugh. One such rule in England stated: 'The speed limit should be four miles per hour. The car should be preceded by a man carrying a red flag.' Another law in Pennsylvania clearly stated that if a horse cart and a motorcar come face to face, then it was the duty of the motorcar driver to give way to the horse cart.

Despite all opposition, the car industry flourished. In 1915, Henry Ford established his assembly line project through which he built a car in 90 minutes flat. He had all the motor parts ready in advance and the workers only had to fit them together. With this technique, Henry Ford could capitalise on the market opportunity. Soon the Ford Company became the world's foremost motorcar company.

Today, there are cars of different shapes, sizes, designs and models. Some run on petrol, others on diesel, some on CNG and there are a few which operate on solar energy.

Motorcars have come a long way and their ride has been a comparatively smooth one.

50 Necktie

I am formal men's wear. I enhance their personality.

The idea of wearing a tie has its origin in the cloth kept by labourers around their neck to wipe the sweat. French soldiers wore a necktie with a knot during the reign of King Louis IV. At that time, it was called a *cravat*. It looked so graceful that the French elite started wearing it. However, the *cravat* went out of fashion after the French Revolution.

In the 19th century, neckties became popular again in Europe. There were many varieties of ties: thin, broad, loosely tied or tight ones. A book named *Neckclothiana alias Tietania* was published in 1818, describing 20 different types of ties.

To tie a knot in a necktie was considered an art at that time. Many people spent a considerable amount of their precious time in trying to tie a perfect knot. To overcome this problem, Sears Roebuck started selling pre-tied neckties. They looked like the ties used by school kids today. They were not very expensive either – six ties for $1.

A tie is made in such a way that no wrinkles show outside even after a knot is tied. In 1920, Jesse Langsdorf made the modern ties we use today. Since then, the necktie has been an integral part of men's formal wear.

In a famous nightclub of Paris, a game concerning neckties became very popular. The woman host of the club would pick out a male guest and ask him to sing or dance. If he refused to comply, the lower portion of his necktie would be severed. People laughed at him. To avoid such an embarrassing situation, most men did as told!

51 Paper

From parchment to papyrus to paper... This is the story of paper in short. Let's elaborate...

In AD 751, China and the Arab countries were engaged in war. Both the sides had some prisoners of war. The Arabs discovered the art of paper making from Chinese prisoners of war. This is how the ancient art of China became known to the world after 600 years.

Ts'ai Lun, a courtier in Chinese Emperor Ho Ti's court, discovered paper in China in the year 105. He made a paste with hemp pieces, rags and bark of the tree. He spread these in thin sheets. After it became dry, Ts'ai found that he could write on it with ink or colour. This was the world's first paper.

However, even before paper was invented, people wrote on papyrus or the thin bark of trees. The word *paper* has been derived from *papyrus*. The Hebrews wrote on the skin of animals. These were called *parchment. Bhoj patras* (thin dried sheets of leaves) were used in India for writing. Many original manuscripts inscribed on *bhoj patras* have been preserved in the National Museum.

Till the end of the 19th century, paper was being made with tree bark, cotton and other items. Slowly, as resources declined, wood powder was used to make paper. On 14 January 1863, a newspaper was printed on such paper in Boston.

This was the beginning.

Today paper is used in writing, printing books, newspaper and even making a paper suit to wear. This suit would have been a great hit had it not dissolved in the rain.

52 Parachute

The first person to think of a parachute was Leonardo da Vinci. He designed it like a pyramid with a square-shaped base. However, the first man to use a parachute was a Frenchman, J.P. Blanchard. In 1785, he put his dog in a basket and attached the basket to a parachute. Then he dropped it from a balloon after it had gained considerable altitude. Having successfully tried a parachute, Blanchard himself descended from a balloon in a parachute. But as he was about to land, the parachute toppled and his leg was fractured.

Another Frenchman J. Garnerin jumped from a height of more than 600 metres from a balloon using a parachute. That was in 1797 in Paris. He had himself designed his parachute.

When his daughter Elisa Garnerin grew up, do you know what profession she selected? She selected a new profession – parachute jumping! She travelled widely in Europe and America and did parachute jumps for fun and exhibits. Great fun indeed! And risky too!

53 Passport

'Those who trouble Potamon on this earth or in the ocean will have to face serious repercussions. They should remember that they will have to face Caesar's wrath.'

Caesar, the Emperor of Rome, gave a memento to Roman philosopher Potamon before he began his voyage to foreign lands. This was the first passport of the world, which gave its owner recognition by his country.

In ancient times, when written documents were not common, travellers were given a ring with the king's picture on it. The Pharaohs of Egypt would give an oval seal to Egyptian travellers. The seal containing their names carried the message that the holder was a representative of the Pharaoh. Any insult to them was treated as a personal insult to the Pharaohs.

In the 11th century, the King of England Canute would give an emblem to pilgrims travelling from their country. This helped them on the way. King Charles II would sign passports personally.

Before the First World War, there was a period when no passport was required to travel to other countries. But the World War changed all that. Once again, passports became mandatory for foreign trips. Since 1914, England and America made it compulsory to paste the photograph of the holder on his passport.

In 1921, the League of Nations changed passports into a 32-page booklet with the photo of the traveller pasted on it. Since then, these passports are used all over the world.

54 Pen

I am the Royal Pen. I lend flair to your handwriting. Remember, how you craved to write with me when you were in the early classes. Let me tell you how I was born.

Hundreds of years ago there were no pens. People would sharpen the thin, hollow sticks of bamboos or use a bird's feather as a pen by filling them with vegetable dyes. In 1809, Joseph Braham made a holder and invented the world's first nib with a feather.

Later he improved these nibs by using horns and tortoise shells. In 1822, he received the patent for these nibs. These nibs were fixed in the holder and each time one had to dip them in ink before writing a few words. Although fountain pens had been given a patent in England in 1809, they did not become popular because their ink leaked badly.

It was only in 1884, almost 75 years later, that L.E. Waterman made fountain pens with a small tank to hold the ink. The ink would flow from this tank to the nib slowly and it did not leak either. These pens were an instant success.

After the Industrial Revolution in 1828, steel pens came into existence. John Mitchell of Birmingham was the first person to make these pens with a machine. It took another hundred years before these pens could be fitted with a steel nib too. In 1926, total steel pens conquered the market and since 1930, detachable nibs increased the life of a pen.

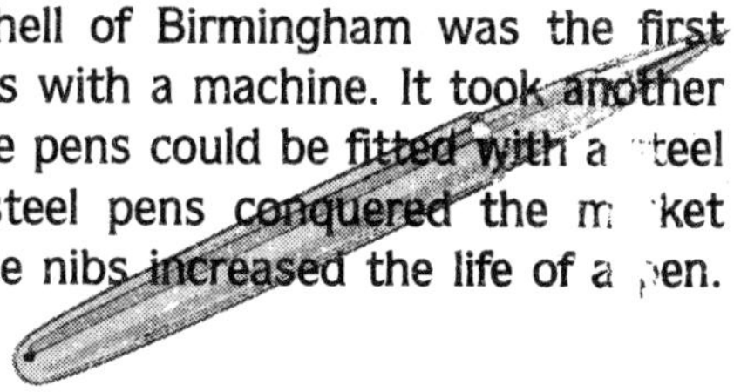

55 Pencil

Hi! I am your friendly pencil. You can write many things with me. Have you ever wondered how I came into existence? I'll tell you the story of my birth.

The story of the pencil goes back to AD 1564. On a dark fateful night, a terrible storm hit England's Burroughdale town. Many trees were uprooted in the storm.

The next day when shepherds went out to check, they found some black stones scattered around the roots of one tree. Taking them to be pieces of coal, they broke these black stones into smaller pieces and tried to use them as fuel. However, this coal would not burn at all. Rather, it soiled their hands wherever they touched those pieces. Then the shepherds realised that it was not coal but wad. Graphite was called wad in those days.

They started marking their sheep with these black pieces of stone. They would pick out long pieces of wad from those black stones and tie a thread around them. This not only made it easier to hold them but also kept their hands clean. This was how the world's first pencil was born. Gradually, the whole of England started using these pencils, since they were very convenient and useful.

In 1565, almost a year later, Konrad Gessner stuck a thin length of graphite in a piece of wood and made drawings with it. This was the ancient form of today's modern pencils. England had abundant stock of wad in Burroughdale, but other European countries were not so lucky. So they started searching for an alternate to wad. The Germans experimented with graphite powder by mixing it with sulphur or gum. They did succeed partially but the quality of these pencils was not very good.

The permanent solution for this problem came about in the times of Napoleon. During the war, England stopped sending graphite to France. Napoleon ordered Nicholas Jacques Conte to look for an alternate to graphite. After much labour and numerous experiments, Conte managed to discover that if graphite powder was mixed with clay, it could be cut into small thin pieces. These pieces could be heated in the kiln to make them stronger.

Around the same time, Joseph Hardmuth of Vienna discovered that the amount of clay mixed with graphite powder determines the softness of the pencil. Even today, pencils are made based on these principles.

In 1950, experiments were conducted using liquid graphite in place of solid graphite. The idea behind these experiments was that the graphite would not break in such pencils and there would be no need to sharpen them either. The fine powder of graphite was made into a thick paste by mixing it with a certain liquid. But these pencils were not very successful because they were not practical. Firstly, their impression was too light. When people put pressure to make it darker, it left such a dark imprint on the paper that it became impossible to rub it off. Finally such pencils went out of use.

Today an ordinary pencil can write 45,000 words or a 35-mile-long line can be drawn with it.

However, one thing is absolutely clear today – there is no LEAD in a lead pencil. In fact, what we consider lead is actually pure graphite. The word *pencil* comes from a Latin word *penicillus*, meaning 'a little tail'. Today there are over 350 kinds of pencils to write on glass, cloth, plastic, film reels and so on.

56 Perfumes

The scent of jasmine, rose and sandal were what enchanted Queen Cleopatra and Noorjahan. Perfumes enhance our spirit. Have you ever wondered how perfumes are made?

Noorjahan, the beloved Queen of Mughal Emperor Jehangir, prepared India's first perfume. Once she and the Emperor were enjoying a boat ride in their palace pond. The queen noticed some oily substance floating above the water. Curious, the queen had it collected separately. For many days, this substance retained its flavour. The queen ordered her servants to separate this oily substance from water. This was called scent or *attar*.

Indian *attars* became famous all over the world. Egypt, Babylon, China, Tibet, Japan, Iran and many other countries would import Indian *attars*.

Cleopatra, the Queen of Egypt, had been also very fond of scents. She had the essence of scented flowers extracted and used them as perfumes.

The advancement in technology and the discovery of crude oil changed the scenario. Now there are scientific methods to prepare perfumes. Scientists have been able to discover artificial scents that are not available naturally. Indian *attar* is still very much in demand since it is famous for its good fragrance also used in certain medicines. The Central Institute of Medicinal and Aromatic Plants has a distillation plant to extract the essence of roses. *Attar* prepared in these plants is pure, and of the best quality.

57 Petrol Pump

A petrol pump is the place where you can buy petrol. Everyone knows about the birth of petrol but what about petrol pumps? Let's share this secret with you...

In the beginning of this century, maintaining a motorcar was a big headache for people because petrol was not easily available. The owner of the car had to approach either an ironsmith or a grocery store to buy petrol.

The owner himself had to fill his can with the petrol, as shopkeepers found it below their dignity to supply petrol through a keep. The main reason behind this attitude was that petrol was an expensive but slow-moving commodity. Even small oil companies too did not give much importance to petrol and placed their stake in kerosene oil instead.

When the number of motorcars increased on the road, people preferred to go directly to the oil companies rather than carry a can to the grocer. Now there were long queues before oil company gates. To avoid this, oil companies opened many petrol pumps at different places. When and where the world's first petrol pump opened is difficult to pinpoint, but in 1912 there was a posh petrol pump in Louisiana with a separate restroom for women. A maid was appointed there to serve cold water to all visitors.

By 1930s, petrol pumps mushroomed everywhere. With so much competition, they used various tactics to attract customers. Some built their pumps in the shape of a Chinese

Pagoda or a Greek temple or a Spanish church. Well dressed, smiling young boys would fill petrol in cars and cleaned windows and windscreens. In one such gimmick, a petrol pump in Maryland hired young bikini-clad girls who supplied petrol to customers.

But as demand increased and the rates of petrol stabilised, such competition died out and, soon, every petrol pump started looking alike just as we see them today.

However, the bygone days seem to be coming back. Self-service petrol pumps are now in vogue. In many developed nations, motorists fill their car tanks themselves and pay at the counter. They are watched through a video screen inside the pump. The whole process is computerised, leaving no chance to cheat.

Fully automatic petrol pumps will soon be operational, wherein the customer would get only as much petrol as the money he deposits in the pump. Although customers may enjoy this novelty, they will surely miss the human touch whereby attendants not only clean the windscreen but also give directions.

58 Photostat (Xerox Machine)

I can make copies of documents. If you could fit in my chamber, I could make a copy of you too!

A small dark room in New York located behind a beauty parlour had two young men working diligently on a table. One of them was Chester F. Carlson. Carlson was a lawyer by profession and always broke. He would deal with many patent cases. This made him think that if he could lay his hands on any such discovery, a patent would then solve all his problems. After much thinking, Carlson decided to build a machine that could make an exact copy of a document. He rented a small darkroom for experiments. Another young man helped him.

One day Carlson wrote 10-22-38 Astoria (22 October 1938) on a glass plate. Then he placed a sulphur-coated metal plate below that glass plate and kept it illuminated with a high voltage bulb for some time. After removing the metal plate, he spread some coloured salt on it. Then he pressed a plain paper on it. His words had been copied on the paper. This was the world's first photocopy. The whole process took not more than a few minutes, whereas rewriting a document took no less than half an hour. Carlson was elated that he had hit the jackpot.

But it was not to be. Carlson went from pillar to post to get his invention acknowledged, but no one seemed interested. Even the famous IBM Company turned him down.

Almost eight years later, in 1940, the Haloid Company in Rochester called Carlson. They evinced interest in his project. Soon photostat machines were made. The market welcomed them with open hands. Now Carlson opened his own company called Xerox Corporation. Its first machine was sold in 1950. Today it is the world's largest manufacturer of photocopy machines.

Xerox machines are much advanced today. The modern Xerox machines are so advanced that besides photocopy, they can also change the size of the text. Coloured photostat machines are also available, which can give you the same colour print in seconds. Life without a photocopy machine seems impossible now.

59 Pin and Safety Pin

Hi, we are your friends! Just be careful while using us and we promise we'll never hurt you.

Pins

Pins were invented in the times of early men. They wore animal skins since clothes had not been invented until then. After wearing the skin, they used pins made of bones or thorns to tie both ends of the skin. With the advent of the Bronze Age, pins were made of bronze. But their design and size was not altered. Even needles for stitching were also made by making a hole in one end of these pins.

Pins are used for various purposes. In the 14th century, the Queen of England levied tax on pins as their demand was at an all-time high. The common man could buy tax-free pins only on January 1 and 2 every year. Ladies would save money around the year to be able to buy sufficient pins to last them for the whole year.

Safety Pins

Many people believe that the safety pin was invented in the 19th century, but this was not so. The safety pin has been used in Rome since the Bronze Age. Safety pins made of bronze have been recovered from ancient tombs in Rome. It is another story that as the Roman Empire declined, safety pins were also lost in the dust of history and anonymity.

Safety pins were rediscovered in the 1840s. Walter Hunt of New York was living in poverty when, one fine day, he folded a simple pin and gave it the shape of a safety pin. He found it very useful. At that time, he was under heavy debt. He went to a merchant and gave him the rights for this pin in exchange for clearing all his debts. All this took merely three hours. And thereafter, the safety pin took the world by storm.

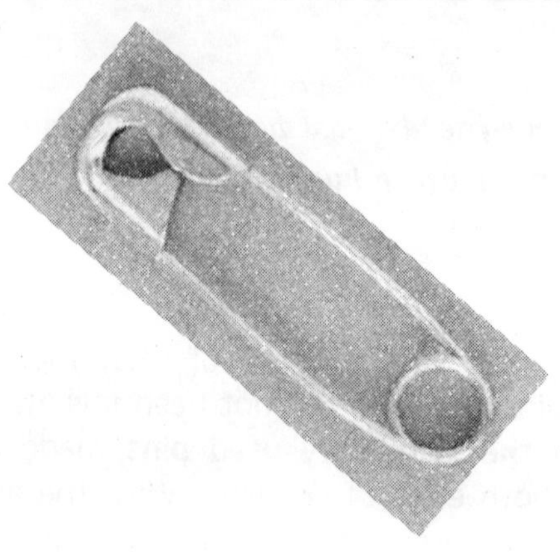

60 Playing Cards

Bored to death with nothing to do? How about a game of cards? No one can defy the magic of the wonderful pack of cards. Let's discover how cards came into existence...

It is not known how, when and where playing cards were invented. Though history does not talk about them, it is believed this tradition began in China. A Chinese Emperor is supposed to have invented them to please his beloved. Slowly this game spread to the entire world.

There was a time when playing cards were thought to be a bad omen. Priests preached against playing cards. It is said that the great explorer Columbus was sailing when suddenly a mighty storm took his ship unawares. All the people on board thought that God was punishing them for playing cards. Quickly, all the sailors on board threw away their cards into the ocean. Luckily their ship anchored at San Salvador safely.

The year was 1942. A war was raging between America and North Africa. One day, as the American army was ready to attack, dense fog hindered their progress. Dwight Eisenhower, who later became the President of USA, was their commander. To kill time, he played cards with his soldiers. He liked the game so much that he continued playing cards even after he became President. He became so efficient in Bridge (a card game) it was claimed that had he entered tournaments he could easily have become world champion.

In ancient times, cards did not have the faces of Jack, the Queen and King. Cards with these faces originated in France. All the four kings had faces of real rulers. King David was the King of Spades. Alexander was the King of Clubs. Julius Caesar was the King of Diamonds and King Charlemagne was the King of Hearts.

Card games are full of superstitions. If the ace of spade falls on the ground or if your elbow touches the deuce of spades, it is considered a bad omen.

Today people gamble all over the world and try to earn money by dubious means. Strange as it may sound, paper currency in North America began with playing cards. Different numbered cards were used as currency of their equivalent number.

61 Postage Stamps

Hey kids, you must have seen me on envelopes! Many people collect me as a hobby.

The postage stamp originated in England in 1680. A man named William Dockwra started a private postal service in London and Westminster. He charged a penny for each letter. The letters were stamped as Penny Post-paid. This stamp was famous as the Dockwra Mark.

James Chalmers and Rowland Hill both claim to be the first to build a self-sticking stamp. Although James Chalmers had made such stamps in 1834, somehow the authorities did not pay any attention to him. Instead, when a committee headed by Rowland Hill proposed the printed, self-sticking postage stamps, it was accepted unanimously. From 6 May 1840, the GPO (General Post Office) in London began selling these stamps. They were priced at one penny for black tickets and two pennies for blue tickets. Many a times, the black stamp of the post office would not be visible on the black stamp, so people reused the black stamp. To overcome this, in 1941 the colour of the black stamp was changed to red.

Till 1854, the postal department had to face major problems since they had to cut each stamp very carefully before selling them. Henry Arthur built a machine in this year, which made tiny holes on the sides of the stamps so that they could be separated from each other easily.

As the number of stamps increased, it became a hobby for many people to collect stamps of different countries. In 1865, this hobby was given a name – *Philately*. Today, collecting old postage stamps is a very profitable business. Very old stamps and the faulty ones too, which are very rare, can be sold for millions.

In 1847, a series of postal stamps in Mauritius had Post Paid printed instead of Post Office by mistake. Today, these stamps are one of the most expensive ones in the world.

62 Pressure Cooker

Do you recognise me? I belong to the 'Digester' family. Today I am known as pressure cooker. Let me tell you my story...

Pressure cookers are based on a simple law of Physics: When the pressure is increased, the boiling point of that liquid also increases.

A French scientist Denis Pepin used this law for inventing pressure cookers. Pepin was working as an assistant to England's famous scientist Sir Robert Boyle. His first successful pressure cooker hit the market in 1672. Pepin called it *Digester.* It could cook even tough food like meat and make it tender in no time. Since it was dangerous to hold steam in an airtight container, as it could burst due to high pressure, Pepin used a safety valve in the pressure cooker, which helped in releasing the extra steam without damaging the container.

Pressure cookers have come a long way. Today there are many varieties of pressure cookers in the market. Pepin's claims of cooking nutritious food in the pressure cookers, while saving fuel and time, proved to be true.

Modern pressure cookers are easier to handle and fitted with a rubber tube called *gasket*, which does not let steam escape. Steam can only be released through the safety valves. Their shapes and sizes have also changed according to requirement. Some modern pressure cookers are portable too.

63 Rope

The Indian Rope Trick is famous all over the world. In this, the magician throws a rope into the air. The rope does not fall but stays suspended and stiff like a stick. An assistant climbs this rope and disappears into thin air. When he does not return even after a long time, the magician climbs the rope holding a knife in his mouth. Bloodstained body parts of the helper start falling on the ground and the magician then climbs down the rope.

Suddenly, the assistant enters the arena from somewhere behind the audience or right in the middle of the crowd – and the applause is deafening. The rope then slinks back to the ground too. No magician in modern times has been able to discover the secret behind this famous Indian Rope Trick. But one thing is certain – this trick cannot be done with the fragile nylon ropes of modern times. Stronger Indian jute and hemp ropes would be required to master this trick.

Jute was discovered long ago during the Stone Age. At that time, the thin branches of a plant were joined to make this. These ropes were used to capture animals or make a net for fishing. As civilisation advanced, the uses of ropes also increased. Ropes were now used in construction, in crossing rivers, building bridges, making charpoys and myriad other tasks.

It is with the aid of ropes that the famous pyramids of Egypt could be built. The formidable Mt Everest would not have been conquered if ropes were not there to help the mountaineers.

Did you know that Asia and Europe were once connected with rope? Three hundred boats were joined together with rope and formed a bridge between Persia and Greece. Xerxes, the King of Persia, attacked Greece across that rope bridge. Unfortunately, he lost the battle and Greek soldiers destroyed the bridge.

The strength of a rope lies in the way it is twisted. To build a thick strong rope, many thin strings are tied together after giving them numerous twists. Earlier, handmade ropes were in use, but after the Industrial Revolution, ropes are also made with machines. To entwine the rope, men once had to walk long distances. The ropes were braided in long sheds. The longest rope till date is about 11.5 miles long. Today different ropes are available for specific jobs.

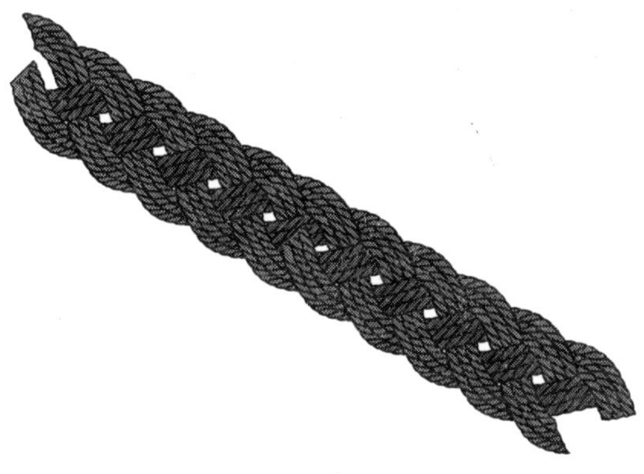

64 Rubber (aka Eraser)

I am a pencil's prime enemy. I can erase a pencil's writing. Some of my cousins are so powerful that they can even erase a pen's writing too. Incidentally, in the UK and many countries once ruled by the British, I'm called a 'rubber'. However, in America, I'm referred to as an 'eraser'. Want to know about my origins?

Consider what Joseph Priestley says about the rubber: "I have seen a new invention recently, which can erase the writings of a pencil without leaving a mark on the paper. I think it will be very useful for painters. This thing can be bought at a shop run by Mr Nairne for three shillings each. He guarantees that one single piece can last for many years."

Mr Priestley described the world's first rubber or eraser in his book published in 1770. It was a great invention of that time. French traveller Magellan discovered the importance of an eraser. In 1752, he submitted his research papers to the French Academy in France. It took 18 years for the Academy to acknowledge the importance of the eraser. But when erasers hit the market in 1770, they sold like hot cakes.

Priestley's book made erasers a famous commodity and Nairne's shop sold them by the dozens. Gradually the market caught up and erasers were easily available in other shops too. However, for a long time it was a small-scale industry. Later, in 1811, a rubber factory was built in Vienna, where erasers were produced on mass scale.

On 30 March 1858, Hyman L. Lipman of Philadelphia (USA) made pencils fitted with a piece of rubber at the back. These became very popular, as they were quite handy. Although such pencils are a common sight these days, way back in the 19th century, acquiring such pencils was considered a great honour.

Erasers in different sizes, colours and fragrances have brought about a revolution in this industry. So take your pick and erase that writing on the wall to create your own history.

65 Safe and Vault

It happened on 4 August 1945 when America dropped an atom bomb on Hiroshima city in Japan. The entire city was destroyed, except for one thing kept just a few metres away from the bomb site. Can you guess what it was? A vault in Teikoku Bank!

As civilisations advanced, people started enjoying luxuries and learnt the value of money. To safeguard their money from thieves and dacoits, they started thinking of innovative ideas like hiding their money in a box or burying it underground or stashing it between walls. Sometimes, people hid their money so imaginatively that even after their death heirs could not locate the whereabouts of the treasure! It is said the Pharaohs of Egypt built pyramids that also held all their treasures, as they believed in life after death. Unfortunately for the heirs, their money was buried along with their dead bodies and could not be used by anyone.

Pharaoh Cheops of Egypt built a pyramid that was 755 ft long and 481 ft high. There were two secret doors to this pyramid. It took 400 years to retrieve all the treasures hidden in this pyramid.

After locks were invented, heavy, iron vaults were fitted with huge locks. When people found that thieves broke such locks, number locks became popular in vaults. But this was also not the best solution because humans being a cautious race, they always kept the number of the lock written somewhere! So these were not very safe either.

Later, fireproof vaults took over from number lock vaults. Since neither the vault nor the lock could be destroyed in the event of fire or theft, they became more popular than other vaults. Now the only way to break open these vaults is via a daylight robbery. Today most banks use these vaults.

The scare of nuclear war has forced many nations to keep priceless documents and other treasures in fireproof vaults so that they are not destroyed in such wars. The National Archives in Washington exhibits some very rare documents from the times of American independence, which are kept in a glass window during the day but placed in a safety vault at night to safeguard them against fire, earthquake, floods and nuclear attacks.

So wake up, my dear! Collect all your precious items and keep them in a vault to safeguard against natural or manmade calamities because that is what vaults are meant for.

66 Safety Razor

Hi! Ever watched your Papa using me for a shave? Here's how my story began...

Razors first came into existence in Egypt. The Egyptians used golden, silver, bronze, and copper razors according to their status. However, very few people shaved. In other countries also, people were afraid of barbers. The big razor never spared anyone. Shaving was like fighting a bloody battle with it. Even a famous general like Napoleon was scared of shaving! For months together people avoided going to a barber for fear of getting hurt with the razor.

By the 18th century, steel razors were made in England and Germany. They had to be honed every now and then and were much safer than their ancestors. Still, a cut here and a cut there was quite common!

In the middle of the 19th century, the safety razor was invented. Jean Jacques Perret went to a barber for a shave. A few days later, he discovered that he had contracted a skin disease. To overcome this problem, he decided to make his own personal razor. He kept the razor blade between two pieces of a carpenter's plane and tried to shave with it – unsuccessfully. This was also a very impractical razor, since the blade had to be sharpened after each shave and the angle of the blade between the planes was disturbed each time.

An American salesman King Camp Gillette tried to correct this problem. Gillette had big dreams. He wanted to be a millionaire. He grabbed this opportunity to fulfil his dream. He thought of placing a steel blade in a holder. Gillette contacted his friend William Nickerson and hired him to make such a razor.

In 1903, Gillette sold the world's first steel blades. He managed to sell 51 razors and 14 dozen blades on the first day itself! People called them safety razors. By the end of the First World War, people started shaving at home and were no more scared of barbers.

Continuing this tradition, Joseph Schick built a battery-operated razor in 1928. It was very convenient while travelling. Since then, there has been no looking back for razors.

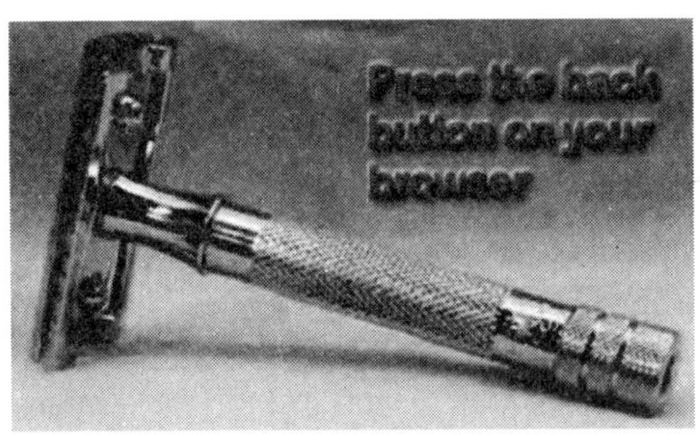

67 Scissor's

Life without scissors seems impossible. Although all it does is cut, it's an irreplaceable cut.

Scissors were considered a mark of death in ancient times. According to Greek legend, three goddesses control a man's destiny. One of them is *Atropos* (meaning heartless), who holds a scissor in her hand. When she cuts the thread of life with that scissor, the man dies. Such stories indicate that scissors were known since ancient times.

Archaeological surveys have shown the presence of scissors in the Bronze Age. These were different from modern scissors. The two blades of ancient scissors were joined in the shape of a 'C' with a spring and were operated using both hands. Later, the Romans made scissors that could be operated with a single hand, quite like our modern-day scissors. They were made of iron or brass. In those days, people used scissors for doing specific jobs.

Scissors became a household item in 1761. The credit for this goes to Robert Hinchliffe of England who made stainless steel scissors which were very strong. Even today, stainless steel scissors are used all over the world.

Today there are scissors of various designs and sizes in the market. Big scissors in the hands of a dress designer can change your style. A small scissor in the hands of a skilled surgeon can give you the gift of life. But a pair of scissors in the hands of a pickpocket can leave you penniless in the middle of the road. Such is the power of the scissor.

68 Screws, Nuts and Bolts

Hi! I'm a screw. Although small, I'm very powerful and important. Let me introduce you to my friends, nuts and bolts. Our story goes like this...

Famous Greek scientist Archimedes discovered the principle of the screw and nuts. During those days, blunt screws were made. Each pair of a screw and the nut was complimentary. It was akin to a lock and key. No other screw could be fitted in that nut! The pair was tied with a thread for fear of losing it. There were no standard norms for making similar screws, nuts and bolts.

In 1904, a fierce fire engulfed the city of Baltimore. Within minutes there was fire everywhere. Fire engines from nine nearby towns could not do anything. You may wonder why. It was simply because their hosepipes could not fit the taps in Baltimore.

During the Second World War, English mechanics could not repair American fighter planes because they did not have the screws and bolts of that size. Today such problems do not exist. There are standard-size screws, nuts and bolts that can be fixed on all machines. During the Industrial Revolution, England's Henry Maudslay tried to standardise screws. After 10 years of intense labour, Maudslay managed to build a machine to make identical screws. This was indeed a big achievement. Nut, bolts and screws, though small in size, are important components of each machine.

69 Sewing Machine

I am a tailor's delight.
Fitting clothes loose or tight
At your service day and night.

In 1841, Bartholomew Thimonnier's 80 sewing machines were working day and night to complete the gigantic task of stitching uniforms for the French Army. The status of the uniforms was definitely much better with 200 stitches per minute with machines, compared to 30 stitches per minute manually. The word soon spread.

Other tailors were now worried, as they sensed doom in their trade. Not everyone possessed a sewing machine. Jealous tailors took the easy way out. They attacked Thimonnier's tailoring shop and broke all his machines. With great difficulty Thimonnier managed to smuggle out one sewing machine and returned to his village. On the way, he demonstrated his machine and earned some money. But this was not sufficient for a livelihood. The next few years were really tough for Thimonnier. He made wooden sewing machines and sold them for $2 each.

In 1845, a trader named Magnin contacted Thimonnier and proposed selling his machines on a large scale. Good days came back for Thimonnier. The machines were picking up in the market when, one day, history repeated itself. Thimonnier had to face the wrath of fellow tailors. They burnt his workshop. Once again, he took to the road.

In the meantime, another man Elias Howe tried to make a sewing machine. He made one such machine in 1846. Unfortunately, he could not sell even a single piece due to its substandard quality.

Isaac Merritt Singer built the world's first practical sewing machine in 1850. By now, many others had also experimented with the idea and managed to build a few different sewing machines. Since Singer's machine was the best by far, many other makers including Elias Howe took Singer to court, seeking compensation for theft of their patent. This event went down in history as the Sewing Machine War. Ultimately, all sat down together and built a patent pool in which everyone joined hands and shared the profits.

In 1856, Singer started selling his sewing machine for $125, which was quite a high price. This was the reason why sales of the machines were low. To overcome this problem, Singer's lawyer Edward Clark, who also had a stake in the business, suggested selling sewing machines on instalments. His idea worked.

Soon Singer's machines were sold all over the world. Today, Singer is the best sewing machine worldwide, and still sells its sewing machines on instalments too.

70 Ship

The world calls me a ship. I've been sailing the seven seas for generations.

The story of the ship begins in ancient times. With the rise of civilisations, man felt the need to move from one place to another. He began with crossing the rivers on wooden logs. Then he joined these logs and made floaters. During 600 BC, sailing boats had been developed. But they were effective only on rivers. Vast oceans needed something stronger and more reliable. People tried to make huge ships with sails but there were many shortcomings, as sailors could not control the direction of the vessel. This depended purely on the direction of the wind. In the absence of a strong breeze, the ship would be left stranded in the middle of the ocean for many days together.

The earliest ship to move with a steam engine was built by John Fitch in 1787. Later, in the 18th century, diesel engines replaced the steam engines. The shape and size of the ship was also altered as time progressed. According to the needs of the people, ships became luxury cruisers. They had all facilities of a five-star hotel like a swimming pool, hospital, playground, radar, radio etc. Huge engines are used to move these ships.

With the advance in technology, ships running on nuclear power were built. The first such ship was built in 1954 in America. In 1959, a Russian ship Lenin was also built on the same lines. America has also built ships that can move across frozen seas. Ships are also useful during times of war – they act as runway for fighter jet planes and other aircrafts.

71 Shirt

If you are asked my story and you don't know it, don't lose your shirt. I will tell you the story of my origin.

If you thought that a shirt is just a cloth to cover your body, you are mistaken. Listen to all these sayings:

- If a pregnant woman wears her husband's shirt throughout the pregnancy, she will give birth to a sweet, healthy child.
- If a person wants to escape an attack, he should wear a shirt that has been stitched by a seven-year-old child overnight.
- To cure an ill person his shirt should be thrown at a crossing.
- If a war widow loves her husband, she should always wear her husband's bloodstained shirt all the time.

Indeed, these were beliefs that prevailed in the Middle Ages, when shirts were the new craze. These shirts were without collars and looked more like kurtas.

The fashion of shirts with collars began in the 16th century. Starched collars were in vogue. Tailors designed different types of collars. Frightfully expensive, frilled and collared shirts were common among the rich. People were so crazy about these shirts that they would go to the extent of selling their property and getting a shirt made! Once a French courtier sold his orchard and had a shirt made from the proceeds. An expensive shirt, indeed!

And although shirts were so expensive, no attempt was made to clean them! People would wear the same shirt for months together and then cast it away! This is the reason people sprayed perfume on their shirts regularly.

By the 18th century, starched shirts came into fashion. Although it sounds like a myth, it was said that these shirts were hard as iron and even a bullet could not penetrate a starched shirt.

In 1820, in a town called Troy in New York, an ironsmith's wife was tired of washing her husband's shirts, as his collars were always dirty. So she found an easier way out. She cut off the collar and kept washing it separately. After ironing, she would button up the collar to the shirt! This started the trend of detachable collars.

By the end of the 19th century, shirts with about 400 types of collars were available in the market! Times change, trends change. Today, sports shirts with an open neck are in vogue. Let's wait and watch what the future holds for us.

72 Shoes

I protect your toes, I am always at your feet.
I may not be important, yet you cannot
walk a single step without me.

In the 15th century, a man in Europe was imprisoned because he wore long-tipped shoes. In those days, shoes denoted the status of a person. The wealthier a person was, the longer the tip of his shoes. There was even a law governing the size of the shoe as per a person's status. A poor man was not allowed to wear long-tipped shoes.

It is said that such shoes were the brainchild of the people of Poland. They had shoes as much as three feet long. These were said to keep witches at bay. Since it was not easy to walk with such shoes on, the tip of these shoes was tied with a chain to any part of the body.

By the end of the 15th century, these shoes were out of fashion. What mattered next was the breadth, not the length, of shoes. There were one-foot broad shoes. When people found them impractical, their size was reduced to 6 inches. This phase did not last very long either.

Next came the fashion of high-heeled shoes. These shoes were especially helpful for short people. The shoes were now fixed with heels of one to two feet high. They were quite uncomfortable. Women lost their balance with such shoes. Many women miscarried after falling down while walking with these shoes. So, such shoes were legally banned.

King Louis XIV started the fad of high-heeled shoes again in the 17th century because he wanted to look taller. Following his example, many courtiers started wearing high-heeled shoes. Later, women also began wearing them. But this time, the heel was not as high as one or two feet high like before. Soon high-heeled shoes became women's wear. Men started wearing flat shoes.

Shoes then gave way to sandals. Leather sandals were popular in Babylon, Athens, and Egypt. They were official footwear in the king's court. The sandal's designs in those days would depend on the social status of the person.

In 1870, wheels were fixed under the shoes much like today's skates. But they were unlike skates. It was not easy to walk with these shoes having knee-length wheels. Soon, these went out of fashion.

Recently, a yoga student of Denmark made a new kind of shoe with a high toe compared to the heel. This is said to be a sure cure for many diseases. The latest in shoes are *disco shoes*, which have a battery-operated bulb on the tip.

Irrespective of size and shape, the fact remains that shoes protect our feet and they will never go out of fashion.

73 Shovel

Hi! I am used in mines. I also come in handy in the garden.

Be it a gold mine or coalmine, shovels are used everywhere. Have you ever wondered while travelling in the air-conditioned compartment of a train that without a shovel it would not have been possible to build the train and its track? Egypt's pyramids, modern skyscrapers or even our humble homes are all built with the help of shovels. In short, no construction can be possible without a shovel.

The world's first shovels were made of wood. Later, only the handle was made of wood, as the rest of the shovel was made of metal. Till the 18th century, England had a monopoly in shovels. In 1774, Captain John Ames of Massachusetts started making iron shovels on a large scale. They were much cheaper than the ones made in England, so their demand rose. Even today, John Ames is the biggest shovel-making company in America.

In those days, each labourer had to carry his own shovel to work! They always kept their shovels safely, as it was a thing of pride for them. However, their work being rough, shovels would break in the course of time and they had to buy a new shovel. This burnt a hole in their pocket. As the trade expanded, some big companies started providing shovels to labourers. This was a good incentive and it helped them find labour at cheaper rates too. This also increased their efficiency. Soon this became the secret of the trade and now, as a rule, all construction/mining companies provide shovels to their labourers.

Today bulldozers have taken over shovels. In big construction works, shovels consume more time, so bulldozers are more feasible. But the fact remains that shovels are still as important as ever.

74 Skates

Zoooooooom! I am a pair of skates, the shoes on wheels – wheeeee!

The world's first skates were made in England in 1760. Joseph Merlin, a violin player, made these skates. To popularise his invention, Merlin performed on stage, playing a violin while dancing on skates. Unfortunately, he fell on an expensive mirror and was hurt. Scared, Merlin never tried the feat again.

About 30 years later, in 1790, skates became popular in Germany and France. In 1818, some ballet dancers performed on snow, dancing with skates. The show was a great success. Soon skates were found everywhere. Skating on snow became all the more popular. By 1860, skating was done purely for adventure and entertainment.

James Plimpton of America first began skating on the floor. He had not been keeping well. So doctors advised him light exercise. Plimpton began to skate on snow. But melting snow created problems. Instead of abandoning skating, Plimpton modified his skates, making them more manoeuvrable, and began skating on the floor. He found he could control himself more easily and balance on the skates by bending left or right with these new skates. To make these skates popular, Plimpton opened many skating rinks in New York. Within three years, skating became America's favourite sport.

Skates were very expensive in those days. Each pair of skates would cost $20, which was beyond the reach of the common man. At the end of the 19th century, skates became more common and cheaper too. Now one and all could afford skates, so everyone started enjoying skating.

Skating is not a risky game today. Of course, there have been incidences when people have broken bones, or even died due to head injuries. But all that is history now. Today, helmets and safety pads protect the skater from these dangers of skating.

75 Skiing

Playing in the ice is like a dream come true. Let's see how skiing was born...

Skiing is a popular sport today. But hundreds of years ago, this was not a sport – it was a profession. Nearly thousand years old skis, belonging to a bygone era have been found in Finland and Sweden.

The closest record of skiing is of AD 950, when the King of Norway, Hakoon hired some men to collect taxes. They had to ski to reach faraway snow-capped mountains to do their job. Later, Poland, Russia, Norway, Finland and Sweden had a special army trained to fight on ice who were skilled skiers.

By 1800, many people had started skiing as a hobby. In 1855, a young man called Thompson, a trained skier, was looking for a job when he chanced upon an advertisement in the employment news – 'Young man wanted to carry mail from California to Genoa. Beware – the two places are 75 miles apart with snow-capped mountains on the way'.

It was surely a dangerous task. But Thompson had full faith in his abilities. He built 9 feet long and 8-inch broad skis. Carrying some snacks and the mail, worth 50 pounds, he started his journey from California to Genoa on skis. Most people believed he would never return. But he was back in California on the fifth day with the return mail. Later, Thompson became a skiing instructor and took active part in organising skiing competitions.

76 Skyscrapers

Strain your neck to look at me. I am as high as the sky. Yes, I am a skyscraper. Curious about my origins? Then read on...

The world's first skyscraper was built in Chicago in 1885. There were 10 floors in this building, which was designed by William Le Baron Jenny. Five years later, another 14-floored high-rise was built in New York. Skyscrapers could only be built because Otis had already invented the lift in 1854. The lift is an essential component of a high-rise building.

The shortage of space and increasing population forced people to favour skyscrapers. As the number of high-rise buildings increased, complications welled up too. Psychiatrists opposed these buildings. They feared these buildings would make residents more prone to psychological disorders. This has been partially true in the sense that people living in a 100-floor building start living in a world of their own. Their home, neighbourhood, marketplace, hotel, recreational centres, lawn, and swimming pool are all in the building itself. They need not venture out of the building. In this way, they start living a secluded life away from others and tend to become more reclusive, even self-centred.

The worst fear in high-rise buildings is regarding fire. An American film *The Towering Inferno* gave a true and graphic account of a raging fire in a skyscraper. When this film was released, the rates of flats in skyscrapers fell sharply. People started moving to bungalows or ground floor flats. However, space compulsions made them backtrack soon and everything was back to "normal".

Skyscrapers enjoy a royal status today. For years, New York's Empire State Building was the world's highest building until Kuala Lumpur's Petronas Twin Towers overtook it; and the latter in turn have been overhauled by Taipei 101, a 101-storeyed building rising 1,670 feet high in Taipei, Taiwan, in 2004.

77 Soap

I am a soap. I am the epitome of cleanliness. I am sure my invigorating story will freshen you up just as my refreshing fragrance freshens up everyone.

The story of soap begins from ancient Rome where washermen built public toilets. Can you link a public toilet with a washerman? I am sure you can't! But hold your breath and read what I have to reveal. These public toilets helped washermen earn their livelihood. The washermen would soak dirty clothes in lye water first; then the clothes were soaked in rotten stinking urine collected in the toilet. Finally, they were rinsed in clean water. No wonder public toilets were of great use to them! During those times, there were no bathing soaps either, so people would scratch their skin with knives to clean dirt and grime!

Soap was probably made in the first century but it came to be used commonly only in the 10th century. Soap was a luxury in those days. Only the rich could afford it. In the 14th century, mass production of soap began in England. In the early years, soap was made by candle-makers because fat and oil are common ingredients in candles and soaps. Big blocks of soap were made and shopkeepers would cut them before these were sold.

Soaps and the world of advertising are seemingly synonymous. The world's first soap *Lifebuoy*, made by Lever Bros in 1918, was marketed as the health soap. Since then, all soap companies have followed suit and the tradition continues till today.

78 Spectacles

I clear your vision. I am your friend for life.

Till the 13th century, no one knew what spectacles were. Short- and far-sightedness were unheard of. In 1250, when French priest Father William Rubruk visited Mongolia, he saw some tribals making two holes in a tortoise shell and fixing two transparent stones in it. Then they would fit the shell on their nose in such a way that the stones would cover the eye. Rubruk was very curious. After making enquiries, he discovered that these stones made one's vision clearer.

After returning to France, Rubruk discussed this with his friend Roger Bacon, who was a scientist. Bacon made convex lenses with the help of those stones. But it was tough to keep them stable on the nose since no frames had been invented then. By 1300, these lenses became very popular.

In the 15th century, with the discovery of the printing press, the older generation realised that they could not read matter printed in small print clearly. These lenses became more useful now. There were no criteria for the thickness of the lens. People would buy any lens that helped them read accurately. In the beginning of the 16th century, concave lenses were also made.

However, famous scientist Benjamin Franklin built the world's first proper spectacles by placing two lenses in a frame. When Franklin was young, he was far-sighted and could read clearly with these glasses. But as time passed, he became shortsighted too. Now he needed two separate spectacles. Franklin found it very difficult to keep shuffling the glasses. So he had the upper portion of the convex lens cut and replaced it with a concave lens. This is how the world's first bifocal specs were made.

Sunglasses are another form of glasses that now protect our eyes from the sun's harmful rays. But when these dark-coloured glasses began their early years in China, judges wore them to hide their emotions. Soon, sunglasses became a rage and a sign of social status. Later, the fashion world adopted sunglasses and today they are very much in vogue.

79 Stapler

I am a part of office stationery. But I have many other uses too. Read on and find out...

The origin of the stapler is not known. But there is evidence that the stapler in its crude form was available in offices in the beginning of the 19th century. However, pins had to be inserted one after the other. So these early staplers were much of a nuisance.

Later, the stapler was improved and a set of 25 pins was made, which could be inserted in the stapler at a time. But these staplers were very heavy and had to be hit with a hammer. Modern staplers can be pressed with the thumb and have come about recently.

Staplers gave various new ideas to people. A powerful staple gun was built and this replaced the nail and hammer. Carpenters, masons, and tailors all use the stapler gun in making tables, chairs, boats, cars, in fixing curtains, sofa covers and building partitions too. Even artists use stapler pins to fix their canvas on the board.

The field of medicine is not too far behind in the use of the stapler either. A Hungarian surgeon built a stapler for use in surgery. Though unsuccessful, it laid the foundation. In the 1870s, American scientists made staplers that could stitch the wound after an operation. Since the staplers were made of stainless steel, there was no fear of any infection in the wound. These staplers were B-shaped, so the flow of blood was also not disturbed.

Finally, staplers have a special use for obese people. For people who cannot control their diet, their stomach is made smaller by inserting a stapler pin in it. When they try to eat more than needed, the stomach stretches and the pin starts hurting and forces them to eat less! Based on acupuncture, another treatment for obesity includes putting a stapler pin in the ear. The patient is asked to scratch the ear whenever he feels hungry. It has been seen that pangs of hunger are considerably reduced with this method.

80 Table

The world's first table was a square piece of stone in a cave on which the caveman kept his belongings. He would eat on that table, sharpen his tools, and did many other things.

As the Stone Age ended, tables changed accordingly. Wooden tables were more handy and easy to make. Initially, they were crude and without polish or artistic carvings. Later, beautiful wooden tables were made. These are now kept in museums and enchant visitors even today.

The most popular use of a table is for eating. Special tables for this purpose were made in the Middle Ages called dining tables. The dining table gave birth to table manners. Licking fingers, making sounds with the knife and fork or while eating were all considered bad table manners and these haven't changed even years later.

Although four-legged tables are an all-time favourite, there have been experiments in which single-legged round tables or oval tables and three-legged round tables and folding tables are worth mentioning. Square or rectangular tables are mostly used in offices and schools, whereas oval or round tables are used for dining purposes or holding special meetings. The famous Round Table Conference in London was a milestone in Indian history.

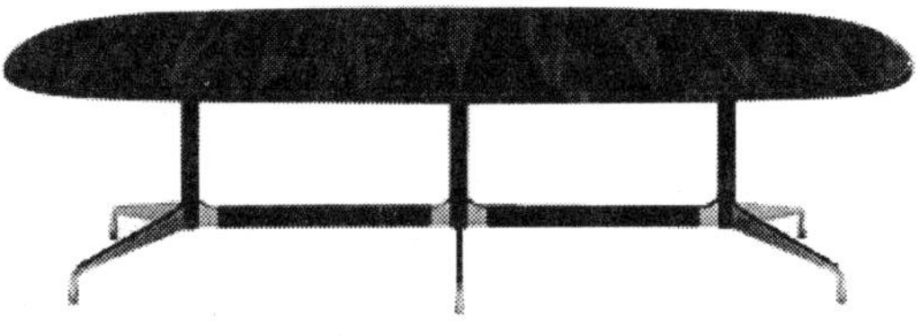

81 Tape Recorder

Hi! I am a tape recorder. I can record, replay and give you the best music of your choice. I am not very old. I was born in the year 1899.

Valdemair Polson of Copenhagen invented this instrument that had two boxes tied with a steel wire or tape. This tape was covered with a magnetic material. To record someone's voice this tape was played. The tape would automatically move from one box to another and record whatever was said through the microphone.

The tape recorder was earlier called *tele-gramophone*. But it was not very successful because the steel tape was not able to give good quality recording. In 1929, German scientist Flumer discovered the magnetic tape. Soon thereafter, the discovery of PVC (poly vinyl chloride) brought about a revolution in this industry and tape recorders were then manufactured on a mass scale. In 1935, an improved version of the tape recorder called the magnetophone was made in Germany.

Since then, tape recorders have never looked back. Today, very sensitive models are available that can even pick up the slightest sound made far away.

82 Taxi

I am a taxi. Earlier I was known as a cab. Here goes my story...

Driving a taxi was a tedious task in the early years because drivers were not given a licence easily. They were also given tough training before their taxi was registered.

The word '*taxi*' has evolved from '*taxi meter*'. Wilhelm Brunn invented this instrument in 1891. Once fitted, this meter could exactly state the distance travelled by the taxi. When Brunn demonstrated his instrument for the first time on the roads of Frankfurt, hundreds of cab drivers were furious. They felt like a tax was imposed upon them. In a fit of anger, they threw Brunn into the river.

In 1907, a rich American, Harry Allen took a cab for a short distance. The rates charged by the driver were very high. This enraged Allen and he brought out his own fleet of cabs, which were fitted with meters. When the common man rode these cabs, he found them hassle-free and much cheaper. Soon, all cabs were fitted with taxi meters and thereafter came to be called *taxis*.

83 Tea

Tea is a refreshing beverage. People all over the world drink tea. Let's hear the story of its origin.

In 2700 BC, Shen Nung was the Emperor of China. He always drank boiled water. One day when the water was being boiled, some twigs and leaves fell into it by mistake. When the Emperor drank that water he immediately called the chef. The chef was frightened but on meeting the Emperor he was amazed to find the ruler in a very good mood. The Emperor rewarded him for preparing the flavoured boiled water. The leaves that fell into the water were tea leaves. This was the world's first cup of tea that the Emperor drank without milk or sugar. This drink was included in the daily diet of the Emperor thereafter. Soon, his courtiers and the whole of China began drinking tea.

But it took 4000 years for tea to reach Europe. The credit for introducing tea to Europe goes to Dutch officers of the East India Company from China. This happened in 1610 when they brought tea to England from China. By 1650, tea was the most popular drink in England. Since tea was imported from China, a huge amount of money was being sent to China in exchange for large imports of tea.

To overcome this problem, the British spread rumours that tea was a harmful drink, so that people would avoid drinking tea and the consumption would come down. A priest Dr Hales even conducted an experiment with piglets and showed that if the pig's tail was dipped in tea for a long time, all its hair fell off.

Yet people were so addicted to tea by now that they would not give up their favourite drink. Even today, the consumption of tea is highest in England among all the countries in the world.

Tea is not so popular in America although the American Revolution was triggered off due to the 342 cartons of tea that were dumped at Boston Harbour on 16 December 1773. Americans are rather fond of their cup of coffee. Serving tea is a tradition in many Asian countries. The beverage is served and made differently in every culture. Japan leads the wagon with its rules for serving tea being as old as 600 years. Drinking tea is more of an occasion in Japan and it is called Ch-no-yu.

Talking about tea reminds one of the kettle in which it is made. A kettle has also been transformed over the years from ones made of clay to ceramic, steel, gold and silver kettles. Different designs of cups and kettles have also flooded the market from time to time, as people fancy innovative designs. The world's greatest revolutions, wars and mutinies have taken place amidst the sips of delicious tea. This probably gave birth to the adage – *a storm in a tea cup.*

84 Telephone

Trriinngg... Trriinngg... Hello, who's calling?

This happened many years ago. In a high-rise building in New York, a man was standing on the windowsill, shouting, "Move aside, I am going to jump from here. I want to commit suicide." Within minutes, the fire brigade and police were there. Newspaper reporters also arrived. People tried to make him see sense, but he refused to budge from the window. Just then, a clever reporter had an idea. He dialled the desperate man's telephone number. The moment his phone rang, the curious man came down from the window to attend his call! Thus his life was saved.

This proves what amazing power a telephone ring possesses. It could bring back someone from the clutches of death. Do you know who was the first person to hear a telephone bell? He was Thomas A. Watson. Watson left his studies at the age of 14 and started working in an electric workshop near his house. This shop helped people who had innovative ideas but lacked the means and expertise. One such young man Alexander Graham Bell was also there. Bell taught phonetics at Boston University. Watson assisted Bell in the invention of the telephone. Watson received the first telephone call from Bell on 10 March 1876. Watson was very surprised at first and could not believe it. In 1877, when Bell established his telephone company, he made Watson a partner in it.

The world's first telephone was displayed at an exhibition on 25 June 1876 but no one seemed interested. However, the whole scenario changed when the King of Brazil, who visited the exhibition, picked up the receiver and exclaimed, "Oh my God, this thing talks!" This was the beginning.

Soon the telephone became a famous invention. The Queen of England asked Bell to display this telephone at her palace. On 14 July 1877, Bell's telephone was shown to the Queen in Canterbury Hall. The Queen liked it immediately and since then the telephone has never looked back.

After the making of the first telephone, it took 39 years before the first telephone service could be started. On 25 January 1915, Bell inaugurated the first telephone line between New York and San Francisco. On the other end was his dear friend Watson once again, who was not surprised to hear his voice this time. The two enjoyed the conversation and this was the beginning of a new era in the field of communication.

85 Television

I am your best friend. I keep you entertained and make your boring day interesting. Do you wish to know how I was invented?

On 2 October 1925, a dishevelled man in shabby clothes climbed down the stairs with hair falling all over his forehead. He forcibly pulled a 15-year-old young boy up the stairs and made him stand before a machine with strong lights focused on him. He himself went into the other room.

A few minutes later, he came back to the room with a worried expression, only to find the boy missing from the spot. He searched the room and found the boy hiding in a corner. The boy was scared. The dishevelled man gave him some coins and assured the young boy about his safety, before bringing him back to the centre of the room. Then he went back to the adjoining room. A few minutes later, the boy's picture appeared on the screen. This was the world's first television picture. The dishevelled man was John Logie Baird and the first TV artiste who posed reluctantly before the camera was William Tanton.

Baird invented the television while he was in sheer poverty and broke during its making. Many a times, he had to sell some items to buy food. At one time, he had lost all hope. Finally his cousins came to his rescue and he succeeded.

But inventing a TV was only half the battle won. England's BBC (British Broadcasting Corporation), which proudly proclaims its best TV shows today, refused to accept Baird's invention at that time. They point-blank turned down Baird, stating that they had no interest in the new invention.

However, the public liked the concept and took the matter to the British Parliament. Finally, on the recommendations of a Parliamentary Committee, TV service began in England but not with Baird's television transmitter but with the Marconi Company's transmitter.

In 1941, Baird built the colour television, which brought him acclaim faster. Now he did not have to face rejection like before.

Today TV is everyone's favourite. However, it is a bane as well as a boon. Just like its name, **tele** is a Greek word meaning *far away* and **vision** is obtained from a Latin word *videre*, which means *to see*, television evokes dual responses from people. Some people like it, while others do not. Sociologists call it a big blow to social life of humans.

But ask any kid, housewife or old man, or those who have time hanging on their hands, and for them television is a God-sent gift.

86 Thermometer

Hi! I am thermoscope. Yes, that is what my inventor Galileo called me. Let me tell you how I was invented...

A thin glass tube was immersed in a bowl full of water. One end with a bulb was dipped into the hot liquid to find out its temperature. On receiving heat from the hot liquid, the air present in the tube would expand and move into the bowl of water. When the bulb was removed from the hot liquid, the water moved up in the glass tube. The height of water in the glass tube measured the temperature of the liquid.

This was the world's first thermometer, invented by the famous scientist Galileo. He called it *thermoscope*. Later, his friend Santorio Santoria placed a scale on the tube so that it was easier to measure the temperature. However, this thermometer had two drawbacks. First of all the level of the water was affected by air pressure. Secondly, there was no standard unit to measure the temperature.

Ferdinand II solved the first problem in the 17th century. He filled the tube with an alcohol-based liquid and sealed the tube. The second problem found a solution in 1724 when Gabriel Daniel Fahrenheit built the Fahrenheit scale with the melting point of water at 32 degrees F and the normal temperature of a human body as 98.4 degree F.

In 1742, Swedish scientist André Celsius suggested another scale for measuring temperatures in which he took the melting point of water at 0 degrees and the boiling point at 100 degrees C. This was called the Celsius scale. Today this scale is used worldwide.

In those days, the use of the thermometer was limited to laboratories or for meteorological surveys. The first thermometer to measure human temperature was built in 1867 in London. After its success, such thermometers became popular all over the world.

Today, there are different types of thermometers – some are disposable, and others are in the form of a blue strip on which the temperature is measured by counting the number of dots highlighted. There are electronic thermometers also, which work on battery. There are digital thermometers that make a beep sound after recording the temperature and displaying it on the monitor.

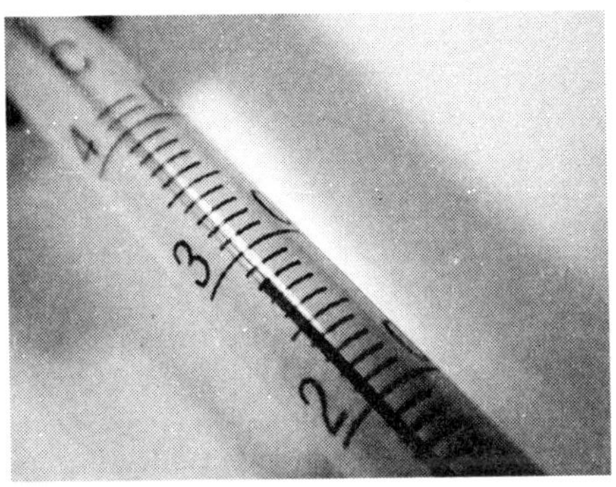

87 Thread

The thread complements the needle. But there are other uses of the thread too. Let's find out these...

According to Greek mythology, 'A man's destiny is like a thread rolled around a spinning wheel. Three goddesses control this thread. Goddess Clotho holds the spinning wheel. Goddess Lachesis starts pulling the thread as soon as the child is born. She keeps measuring the length. At the time of death Goddess Atropos cuts the thread with a scissor.'

This story is proof enough that the thread was used in olden times also. Till the 19th century, thread was made on a spinning wheel, which was manually operated. Europeans preferred the silk thread. Handloom was considered low grade and mostly used in India and other neighbouring countries. At the time of the Anglo-French war, Napoleon had stopped sending silk thread to England. Out of necessity, the English started using cotton thread. But they made it so fine that it gave the silken look.

It was John Mercer who discovered something very important, which changed the life of cloth. Mercer worked in a dyeing shop in London. One day, Mercer observed that if cotton cloth was dipped in caustic soda for some time, the thread became thick and strong and could be easily dyed. This gave birth to *mercerised* cloth. Although Mercer was not a scientist, England's Royal Society felicitated him in 1852 by making him their fellow member for the discovery of mercerised cloth.

However, there are still some more miles to go before thread can be said to be perfect. Even after many unsuccessful attempts, no one has been able to make a thread as strong as a spider's web. Thread woven by a spider is so thin, yet so strong. It has elasticity too. Even the fiercest of storms cannot harm a spider's web. If we could make such a thread one day, it would be the ultimate in a spinning tale.

88 Toothbrush

I am the best friend of your teeth. Thanks to me, I give your face a sparkling smile. I am a toothbrush.

The toothbrush was made in 1780 at Newgate prison in England. A prisoner named William Addis was kept there for starting riots in the city. Like all other prisoners, he too faced a problem in cleaning his teeth. Either they used their fingers or a soft stem. One day, Addis picked up a long bone of an animal and made tiny holes on one side. He fixed hard hair of some animal in those holes. And lo, the world's first toothbrush had been made!

Gradually, everyone started using these toothbrushes, including the prison staff. Addis had hit the jackpot. As soon as he was freed from prison, he started the business of making toothbrushes, which flourished immensely. Instead of bones, he now used metal or wood to make toothbrushes.

In 1938, after the discovery of plastic and nylon, the first toothbrush with nylon bristles was made. Since then, toothbrushes with different sizes, styles and hardness are available in the market. Today even auto-operated toothbrushes are available.

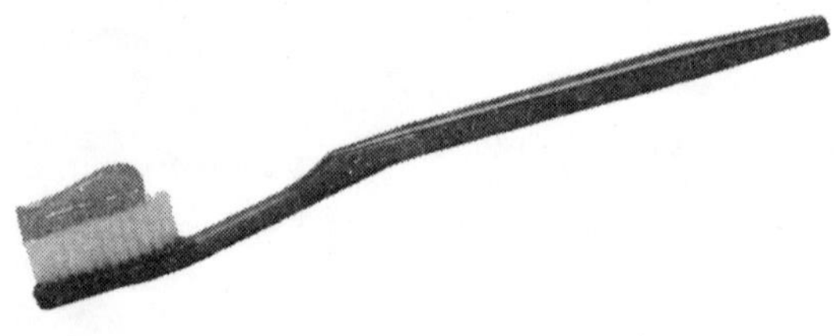

89 Toothpaste

I am the humble toothpaste. But I am very important for your teeth. Let me begin the story with my ancestors.

Since times immemorial, people have been cleaning their teeth with all sorts of things. Hold your breath, while I tell you about some of them. Hippocrates suggested the ash of hare's skull mixed with the ash of three rats. Another teeth cleaning powder was made with burnt eggshells. A special mix of crab's eyes, bat's dropping and iron rust were said to be the best toothpowder, which kept teeth strong and healthy.

It is amazing that people have been using all these things blindly just to keep their teeth sparkling white and enjoy fresh breath. People in ancient Rome rubbed the above substances on their teeth with the help of a muslin cloth. Their main aim was not to keep their teeth clean but to keep them white and sparkling because it added to their glamour.

At the turn of the 19th century, famous scientist Louis Pasteur discovered bacteria. Now people realised that tooth decay occurred due to the presence of bacteria in their mouth. So it is equally important to guard teeth against bacteria, besides keeping them sparkling clean. This is how the modern toothpaste was invented.

Initially the chemicals used in toothpaste were more harmful than helpful. Hexachlorophene contained in the toothpaste, to kill bacteria was also harmful for children. Similarly, chloroform was added for fresh breath and cyclamate was used to give toothpaste a sweet taste. Both chemicals are considered carcinogenous (cancer causing). Apart from this, the tube containing toothpaste was also made of lead. With time, this lead would dissolve in the toothpaste and enter our body.

However, scientists kept eliminating these harmful chemicals and improved the quality of toothpaste. Today's toothpastes are the best, as they ensure teeth are white and sparkling and also remain germ free. They pose no danger to people's health either.

90 Toothpick

The toothpick seems such an insignificant article. Yet, when something gets stuck in our tooth, nothing else helps except the toothpick. History reveals that toothpicks were invented when man started eating meat. Then he felt the need for a toothpick to clean pieces of meat stuck between his teeth.

In 3000 BC, the people of a Sumerian tribe used thin, golden sticks as toothpicks. These toothpicks were attached with their rings. In ancient Rome, people used any bird's feather (except the eagle's) or the thin stem of a tree. Some wise men advised using porcupine's quills. There were many beliefs associated with toothpicks.

Muslims carried a toothpick to ward away evil spirits. In some temples in Japan, toothpicks made of different woods were distributed among disciples. It was believed that they could cure toothche. Toothpicks made from the stem of willow (in spring), cherry (in summers), chestnut (in autumn) and orange (in winters) were commonly distributed.

By the 19th century, some bold merchants commercially produced toothpicks and sold them in the open market. Initially they were not in much demand. So these merchants thought of a strategy. They would send their representatives to expensive restaurants. Here they would ask for a toothpick after the meal. Since restaurants did not keep toothpicks, these representatives would create a commotion. Many restaurants fell for the trick and ordered packs of toothpicks. This is how the toothpick market initially flourished! Even today, there is no restaurant in the world where you will not find a toothpick.

91 Towel

I clean your hands. I keep you dry. I am a towel.

Till the 19th century, towels were not very popular. People dried their hands in the air or wiped them on their clothes! It is believed that most Europeans hardly took bath, so they did not require towels! However the rich would bathe everyday, using a special cloth as a towel. This cloth was very expensive, though. Therefore, a towel was considered a luxury in those days.

The first proper towel in the world was made in 1841 in France from silk cloth. In 1851, Samuel Holt made cotton towels and displayed them at an exhibition held in the Crystal Palace, London. When Queen Victoria of England visited this exhibition, she was bowled over by the towel. She awarded a gold medal to Holt and ordered six-dozen towels for the Royal Palace.

Soon, these towels were household items because they were cheap and easy to use. In 1863, Holt moved to America, where he set up his towel factory in Paterson, New Jersey. Soon his towels were sold all over America. For many years, towels were white in colour because bathrooms were also white. In 1925, coloured towels hit the market.

Rich and influential people did not like the idea of common man using the same towel they used. To maintain their status, they started printing their monogram or logo on towels. Milton Weigler of New York started making monogrammed towels. Soon, Weigler towels were all over the world. In 1970, King Khalid of Saudi Arabia bought some towels from Weigler at the exorbitant rate of $100 per piece. The King's royal logo was printed on these towels.

Today's modern bathrooms proudly display blowers to dry hands. It seems we have come full circle because before towels were made, people dried their hands in the air.

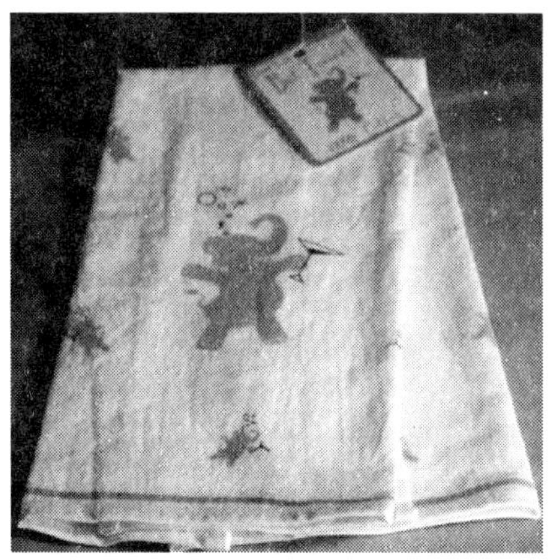

92 Traffic Signal

STOP! says the red light. GO! says the green...

The world's first traffic signal was set up in England in 1868 on George Street and Bridge Street crossing. Petrol lamps were used for illuminating it. The lights were fixed on a seven-meter high pole. A traffic policeman would place a red or green glass before the light with the help of a lever. Unfortunately, the gas cylinder exploded after two months, killing the policeman. This ended the world's first traffic signal.

Electric traffic signals were first set up in 1920 at Woodword Avenue and Michigan Avenue crossing in Detroit City, America. This was one of the most congested crossings at that time. This signal was like our modern signal having red, yellow and green lights, which would blink one after the other at periodic intervals.

Today, traffic signals have also been computerised. There are electronic sensors at different places, which set the blinking time of the traffic light according to traffic density.

Here is an interesting story about traffic lights. In 1950, the clever Mayor of Ludowici, Georgia used traffic lights to earn money. On one of Ludowici's major crossings, he decreased the timing of the green light to 16 seconds. Since the time was too less, most drivers crossed the red light. They were quickly nabbed and made to pay the fine of $15. Soon the mayor had earned $50,000 in a year, which was four times Ludowici's annual budget.

93 Train

Hi! I am black, and I emit smoke from my chimney. You love me because I take you to distant places.

Its smoke will kill the birds.' 'Its sparks will burn our fields.' 'Pregnant women will have a miscarriage.' These were the predictions made by people when in the year 1830 world's first train began its journey from Liverpool to Manchester. Today the whole world knows that all these predictions were wrong.

A train journey in those times was not very comfortable. There were only three to four coaches behind the engine with a capacity of not more than six passengers in each coach. There were wooden benches to sit and the coaches were open from one side. Passengers had to bear the cold wind along with dust and smoke.

To overcome these problems, rich families preferred to travel in their own private coaches, which could be added behind the ordinary bogies of the train whenever they decided to travel. The private coaches had tables, chairs, comfortable beds, crockery, food, wine and all other paraphernalia. Some coaches had a proper kitchen to provide them fresh food too. An opera artiste even had a silver tub in the bathroom of her coach.

Though most leaders in Europe travelled in luxury, some great leaders did not like this discrimination. American President Abraham Lincoln had a special luxury coach at his disposal that he never used. He always travelled by ordinary coaches of the train. Mahatma Gandhi too always travelled in the third class coach of a train and set an example for his countrymen.

Coming back to the problems faced by early passengers during a train journey, apart from discomfort inside the bogies, there were other discomforts too. When coal in the engine ran out, passengers had to fetch wood from the forest! Sometimes, they had to fetch water to cool the engine. The time of a train's departure was fixed but the arrival time was never written in the timetable.

The Orient Express was the world's most famous train. This train was started in 1883 from Paris to Istanbul. It was considered a great honour to travel by this train. The world's great leaders travelled in this train. Many pacts were signed on it. This train has also been mentioned in films and books. To be the Chief Officer of this train one had to be proficient in at least six languages.

King Boris of Bulgaria loved the adventure of travelling in this train. As soon as this train would enter Bulgaria, the King would board it and go straight to the engine. He loved driving this train. The engineers had no say in this. But fortunately, the train was never late for its destination. The King certainly knew his job well.

However, the arrival of aeroplanes overshadowed train journeys. All the same, the thrill of travelling in a train can never be matched by short journeys of air travel.

94 Typewriter

Hi! I am a typewriter. I help you write neatly, easily and quickly.

The typewriter was invented in 1808. Pellegrine Turi of Italy gifted the world's first typewriter to his friend, Countess Carolina Fantoni. The countess would type letters on it and send them to friends. Even today 16 such typewriters exist at a museum in Italy. These typewriters have 23 Italian alphabets and four punctuation marks.

After this, as many as 51 people tried their hand at manufacturing typewriters but there was always some technical problem or the other. Finally, in 1873, Christopher Latham Sholes built the first commercial typewriter, which has been in use ever since. Today's keyboard is an improved version of Sholes' typewriter.

James Densmore commercialised Sholes' typewriter and opened his first shop in the same year at Hanover Street in New York. Each typewriter was priced at $125. This typewriter had two drawbacks: it had all capital letters only and the typist could not read the words while typing. These were soon rectified. Now Densmore signed an agreement with a big company called Remington, which is still a leader in the world of typewriters. Remington have sold typewriters to many prominent personalities, including famous author Mark Twain whose book *Life on the Mississippi* holds the honour of being the first typed manuscript in the world.

With the advancement in technology, today different kinds of typewriters are available: manual typewriters, electric typewriters and electronic typewriters with memory. However, the advent of the computer keyboard has dented the sales of typewriters.

95 Umbrella

Rain, rain go away,
Come again another day.
Little Johnny wants to play. ...

Of course, if little Johnny had carried an umbrella, this problem may not have arisen. There was a time when a man using an umbrella was frowned upon. *'Be man enough and brave the rough weather'* went the adage. Umbrellas were considered a sign of female vanity or weakness. Although women continued using umbrellas till the 16th century, umbrellas were not much in demand.

It was only when priests started using umbrellas in the 16th century did society accept them. But the Pope did not want the common man to use an umbrella, so he passed an order that no one was allowed to use an umbrella without his permission! This order was for men and women alike. Despite this ban, women in London could not give up the charm of an umbrella. Soon the streets of London were full of colourful umbrellas.

Jonah Hanway (1712–1786) of England began the tradition of men using umbrella. Men preferred black- umbrellas. Soon they started using umbrellas without hesitation.

The story of the umbrella dates back to 1100 BC in China, where the world's first umbrella was invented. Egypt also boasts of its ancient umbrellas. These umbrellas were not used as a shield from the sun and rain but were a part of the King's throne. In ancient times, the umbrella was a sign of prestige. The carvings of Egypt's famous pharaohs show a big umbrella above their heads. It is said that these umbrellas were the source of their divine powers. They were made of gold studded with pearls and diamonds.

The Emperor of Japan had a red umbrella as his symbol. The Emperor would not step out of his palace without this umbrella.

In India also, the king's *chhatra* (umbrella) held significance. His commander's *chhatra* (umbrella) held aloft was a sign of victory. If an army's commander returned without his *chhatra*, it meant defeat in the battle for the country.

Today the umbrella is more a utility item rather than a symbol of prestige. People around the world use umbrellas, irrespective of their caste or status.

96 Underground Railway

In England they call me tube, in India they call me Metro Rail, in America they call me the Subway.

In 1912, a tunnel was being dug in the city of New York to lay the railway track for the Metro Rail. During the course of digging, labourers found a tunnel already existed there along with railway tracks and a station. Even a mini train was found standing on the tracks! Who did this belong to? To begin the story of the underground railway, let's go back some 150 years.

In 1860, Alfred Ely Beach was sitting in his office located in the busy Manhattan Street and watching the heavy traffic on the road. Suddenly it struck Beach that if traffic kept increasing like this, the day was not very far when they would have no space to move about. So he decided to find a solution to this problem. Building a railway track on the road was neither feasible nor practical. So Beach thought of building an underground railway station. He submitted a proposal for the underground railway to the authorities but did not get clearance because of political indifference.

So, the clever man that he was, Beach thought of another alternative. He took permission for digging a tunnel and started the digging work at night. After hard labour of 58 nights, he had a 312-foot long tunnel with 9 feet diameter. A station was built. The waiting room was decorated with flowers, fountains etc. A train engine was built with a seating capacity of 20 passengers. Soon the news was out.

But political hurdles were still a major obstacle for the underground railway. It took Beach so long to obtain permissions that financers started backing out. Due to lack of money, Beach had no alternate but to shelve the project. The subway discovered in 1912 was Beach's underground railway station!

By the 1930s, many countries had built Metro trains. The major problem faced by all these projects was building the tunnel. It was a tough and risky affair. Many workers died during each project. But that was not all. There were other problems also. For instance, in Rome, during the digging of the tunnel many artifacts of historical importance were found. Being a city of great historical importance, Rome had passed a resolution that such things should not be touched without the permission of the Archaeological Survey department. The process was slow. Hence the progress of the subway also suffered in Rome.

The Germans faced a worse problem when digging a tunnel in Munich city to build a Metro station. The diggers had to be very careful at each step because many bombs from World War II were still lying scattered under the soil. One careless move and the bombs could go off claiming many lives.

In India, the first Metro Rail was built in Kolkata during the British rule. It is still going strong. The Delhi Metro is largely an overground railway with certain underground sections.

97 Vaccination

Vaccinations give us lifelong immunity towards fatal diseases like polio, smallpox, tetanus etc. Let's see how vaccines came into existence.

This is a long story which began in the 18th century. It started at a time when smallpox had hit the world as an epidemic. Dr Edward Jenner of England felt that people who worked with cows seemed to be less prone to smallpox. He tried to test his findings.

On 14 May 1796, Dr Jenner gave an injection to an eight-year-old boy James Phipps. The injection was filled with a watery substance obtained from a boil of a smallpox patient. The boy showed some symptoms of smallpox and recovered. About a month and a half later, he injected a stronger dose into the boy again. Jenner was pleasantly surprised when he found the boy hale and hearty with no symptoms of smallpox. Dr Jenner concluded that since the boy had developed immunity after the first shot, his body fought against the disease after the second shot and the disease could not affect him.

This was the world's first *vaccine*, so called because *vecca* in Latin means *cow*. This vaccine was used on many patients successfully. Later, Louis Pasteur also developed a vaccine for rabies and saved many lives.

Before long, vaccines for typhoid, whooping cough, tetanus, cholera etc. were discovered. Initially, people were reluctant to take vaccine shots. They would not join the army for this reason alone because it was compulsory for all army personnel to take a full course of vaccination. But soon people realised its value when the whole world was reeling under an epidemic and army personnel were spared.

Since then, vaccination has been made compulsory for small children. Vaccination is a sure shot way of curbing dreaded diseases.

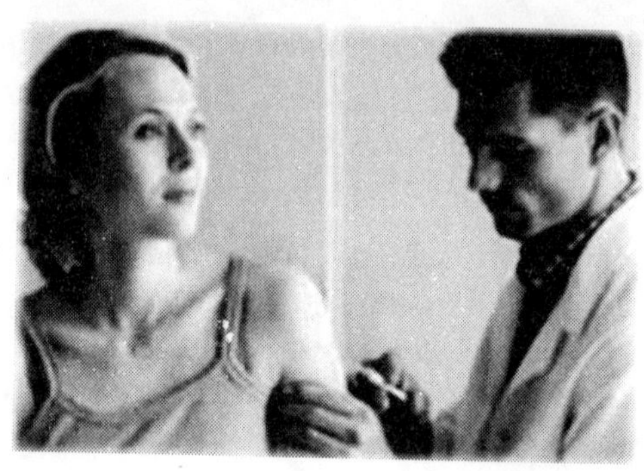

98 Washing Machine

I am your mother's right hand.
I clean your clothes and help you look neat and tidy.
I'm sure you would like to hear my story...

The year – 1941, the Second World War was in full swing. A news item in *The New York Times* amazed the whole world: 'Nazis defeated by Washing Machines.'

Actually what had happened was that the Nazis had overrun Holland and confiscated all butter-making machines from the people. This hit their business hard. But the Dutch were not the ones to give up so easily. They started using their washing machines as butter-making machines and life returned to normal for them!

The usefulness of washing machines extends to what limits can be seen with this example. People of Johnston Island (located 700 miles away from Hawaii) relished octopus meat. But it was very difficult for them to cut it into smaller pieces since it is very hard. When washing machines arrived on the island, they used them to chop octopus meat.

Rodger Rodgerson of England built the world's first washing machine in 1780. A second machine built by Nathaniel Briggs of America in 1797 followed this. Unfortunately, both machines proved a failure. Vigorous efforts continued and, by 1875, more than 2000 models of washing machines had come out and been discarded too. Even their attractive names like Housewife Darling, Pastime, Women Friend, Kill

Kara, Hummer etc. did nothing to attract women. The main reason for their failure was that clothes still had to be rinsed and dried by hand. This was unacceptable to women, as they wanted a machine that did not soil their hands.

In 1922, Howard Snyder built just such a machine. It was automatic and had a circular plate at the bottom, which was attached to four standing blades. This was the basic model of the washing machine, which is used even today. This became immensely popular and improved versions are now available in the market.

Washing machines are not only a boon for women but useful for men too, who are rarely comfortable with washing clothes. It so happened that in 1849 thousands of people went to California in search of gold. There were no women among them. The big problem of washing clothes came up. Do you know what happened? They had to send their clothes to China for laundry!

99 Weighing Machine

My name comes from a Latin word Bilanx. *meaning 'two parts'. The story of my birth goes like this....*

The balance beam was first built when man tried to carry heavy loads on his shoulders. He found that if he distributed the weight equally between the two parts and carried it on a beam, the weight seemed much less. Though the concept was totally different, one thing became clear – when both parts of the balance beam had an equal load, the beam stayed erect. This gave rise to a balance beam that was used for measurements.

A verse from *The Book of the Dead* of ancient Egypt states: '...The heart of the dead will be ascertained on a balance beam. One of its pans will have a truth feather and the other will contain the heart. If both the pans are equal, it means that the heart is pure. God Osiris shall welcome such a human into heaven. But if it is the other way around, the sinner shall be consumed by a demon present there...'

The passage indicates that the balance beam was known in ancient times also. In the early days, since trade was via the barter system, the balance beam was not required for trading and only goldsmiths used balance beams. As trade links spread around the world, a standard unit to measure things was needed. Thus, the balance beam came into use.

In the 16th century, Dr Santorio of Italy built the world's first weighing machine to weigh his patients for diagnosis. By the end of this century the balance beam became quite common. The marketplace, homes, and offices... everyone started using the balance beam. Today computerised machines have been invented that have replaced the manual balance beam.

100 X-rays

I am no James Bond,
But I can see beyond,
I can peep inside you,
Like water in the pond.

'X-rays have arrived.' 'They can take pictures of the soul also.' 'Soon X-ray specs will be in the market, which can see through your clothes.' 'They can even see through walls.' 'It is morally wrong. X-rays should be banned. They are a threat to our privacy.' Such rumours did the rounds when Wilhelm Roentgen discovered X-rays. Strangely, these rays were first demonstrated before children and not before a panel of scientists.

One evening Wilhelm Roentgen threw a party for children and held a magic show. He placed a cathode ray tube and a square piece of cardboard on the table. A closed wooden box was kept between them. He asked the children, "What is in the box?" How could they tell since they did not know? Then Roentgen switched off the light and put on the cathode ray tube. Immediately there were shadows dancing on the cardboard. These were shadows of the weights kept in the wooden box. The children did not know that these shadows were made with X-rays, which were to become a very useful tool in medical science in the years to come.

Not just medical science, X-rays have also been helpful in discovering the past. In 1960, when X-rays of Egyptian mummies were taken, amazing facts were discovered that changed history. Some dates had to be revised. Many dynasties were changed too. Take the example of the mummy of Tutankhamen. The X-ray showed that its skull was broken from behind, as if someone had hit it hard. This clearly indicated a murder. Queen Makare's mummy was buried with a small mummy, which was till now thought to be her newborn baby. But an X-ray discovered the smaller mummy to be that of a baboon and not her infant.

Why are X-rays called so? It is common knowledge in science and maths that any unknown factor is termed 'X'. Since these rays were not defined at that time, they were called X-rays. Thereafter, the name X-rays stuck forever.

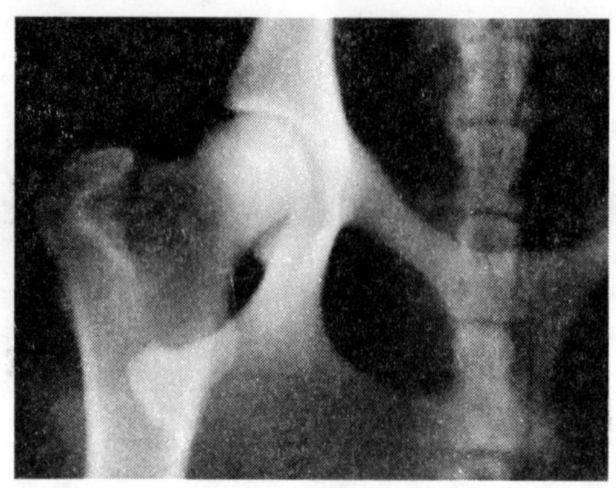

101 Zipper

Yesterday's fasteners are today's zippers. Here's how my story goes...

Whitecomb Judson discovered the zip in 1891. The early zippers were a bunch of hooks that had to be fixed among each other manually. In 1905, after some modifications, they came to be called *C-curity*. Still the zips were not very popular among the masses. The company manufacturing zips was soon on the verge of bankruptcy. In a final attempt, this company hired a Swedish engineer Gideon Sunback to build a cheaper, easy-to-operate fastener. In 1912, Sunback managed to build one such fastener. This is today's zipper.

To prove its worth and longevity, at an exhibition held in Wembley in 1924, a big zipper was placed at a big zip stand. All visitors were free to move it up or down. By the end of the exhibition, this zipper had been opened and closed by more than three million people. It was still running smoothly without any trouble. This inspired confidence in the people that it was an ever-lasting product. The results were outstanding.

In 1927, the company received a bulk order of zips to be fitted in sport suits. Slowly it spread to ladies dresses and gents clothes too. Today almost all dresses are fitted with zips.

But how did the fastener become the zipper? In 1923, the B.F. Goodrich Company built long shoes to be worn during monsoons. It was not feasible to put laces in these shoes, so fasteners were fitted instead. The shoes were sold under the brand name *Mystic Boots*, but sales were not very encouraging. To boost sales, the company renamed these boots 'Zipper'. Whether this increased the sales of the boots is not known but the name stuck and yesterday's fastener became today's zipper.